Springer Biographies

More information about this series at http://www.springer.com/series/13617

Martin Williams

Nile Waters, Saharan Sands

Adventures of a Geomorphologist at Large

Martin Williams
University of Adelaide
Adelaide, SA
Australia

Springer Biographies
ISBN 978-3-319-25443-2 ISBN 978-3-319-25445-6 (eBook)
DOI 10.1007/978-3-319-25445-6

Library of Congress Control Number: 2015950923

Springer Cham Heidelberg New York Dordrecht London

Cover photos: Author and camel, Azaouak valley, Niger (*front*); Tisisat Falls, Blue Nile; Eastern Sahara desert west of the Nile, northern Sudan (*back*)

Printed on acid-free paper

Springer International Publishing AG Switzerland is part of Springer Science+Business Media (www.springer.com)

In memory of my parents, who confronted life's vicissitudes with equanimity, fortitude and humour, and who nurtured my curiosity about the world we live in. This curiosity was the catalyst for my abiding love of the remote and sparsely populated arid regions of our planet.

Preface

Over the years, many friends and colleagues from different parts of the world have suggested that I should put pen to paper and describe in simple, non-technical terms the adventures I have experienced during my work as an earth scientist in some of the remoter parts of the arid and semi-arid world. My aim in writing this concise account of some of the work I have been involved in over the past 50 years is to try and convey for the non-specialist something of the excitement and fun involved in fieldwork in the drier regions of the world. My studies of the soils, landforms and the recent geological history of arid and semi-arid regions have taken me to some remarkable places in Africa, Asia, Australia and the Middle East. Not only are the landscapes themselves often stunningly beautiful, but I have always found the contact with people from quite different backgrounds and cultures an enriching experience. I have been fortunate in that my work has taken me to places far off the beaten track, whether it be the rugged mountains of Ethiopia and northern China, the sandy deserts of the Sahara and Rajasthan, or the great river valleys of Somalia, central India and the Nile. The chapters that follow are not intended to form a coherent chronological narrative, although they do appear in rough chronological order. They should rather be viewed as vignettes or brief evocative descriptions, much as in the discursive tradition of the wandering Irish storytellers or *shenachies*. Acting on the principle that it is not necessary to be solemn to be serious, my aim is to entertain as well as to instruct. If these tales encourage younger people to carry out fieldwork for themselves, then I shall be amply rewarded.

Acknowledgments

My special thanks go to Frances Williams, who drew the map figures with skill and close attention to detail. I am grateful to Rebecca Atkins, Alec Atkins, Mark Busse, David Hall, Professor Emi Ito, Pascal Lluch, Professor Colin Murray-Wallace, Neil Munro, Dr Isabel Vilanova, Patrick Williams and Allyson Williams for their constructive and encouraging comments on draft chapters. I thank Juliana Pitanguy and Meertinus Faber of Springer for constructive suggestions and encouragement. The debt I owe to my field companions will be evident to any reader, but special mention must be made of Lieutenant Colonel David Hall RE, and my late close friends and colleagues Dr Bob Story, Dr Don Adamson, Professor Desmond Clark, Professor Mike Talbot and Dr Françoise Gasse for their sterling qualities of imagination, perseverance and humour. I remain the better for having known them.

Contents

List of Figures

Chapter 1
Early Days

August 4, 1914. My Irish grandfather Joseph Toye (whom I will henceforth call Joe) was having his whiskers trimmed by the English barber (married to the local Welsh post-mistress) in a little coal-mining town in South Wales. Joe was the local police sergeant and before joining the police he had served ten years in the Royal Marines. The radio was on and Joe heard the British Prime Minister Herbert Asquith announce that as a result of the German occupation of Belgium and their failure to withdraw, Britain was now at war with Germany. At that moment the barber leaned over and, knowing that Joe was Irish, assumed he was anti-British and pro-German, and whispered in his ear: 'Now is Ireland's chance!'

M. Williams, *Nile Waters, Saharan Sands*, Springer Biographies,
DOI 10.1007/978-3-319-25445-6_1

He had misjudged his man. Joe said nothing but on crossing the street and entering his home said to Ellen, his Irish wife and my grandmother: 'The man's a bloody traitor, and I'm going to arrest him!' There were two snags to achieving this plan. First, my mother who was nearly three at the time, immediately rushed across the street and piped up with childish glee to one of his little girls, thinking it was some sort of game: 'My daddy's coming to arrest your daddy!' Second, in order to arrest someone on suspicion, Joe had to send a telegram to his superior officers in London, which meant going through the post-office run by the barber's wife. The barber was in fact a German spy. He panicked and caught the first available train to Cardiff, as Joe had anticipated. Joe arrested him on the train. The man was a 'sleeper' named Otto Krüger, in charge of the secret German sabotage network set up years earlier with the aim of incapacitating the coal-mines, tin-plate factories and dockyards of Wales once war began. Krüger had fled with his coded notebooks, enabling the British intelligence service to wind up the entire network within the first few days of the war.

The surname Toye is not an Irish name, as Joe discovered to his intense chagrin one day researching his family tree. It was the surname of an East Anglian yeoman who had deserted from Oliver Cromwell's army during the latter's genocidal campaign in Ireland in 1649–50. He had fled to the mountains of northwest Ireland, married a local girl and become 'more Irish than the Irish' to use my mother's phrase.

My mother Ethna was born in Barry, a seaside town in South Wales. Joe insisted that she learn Welsh 'so that you know what the beggars are saying!' Her Welsh teacher, the exuberant Gladdie, was one of my father's seven sisters. Dad's family lived in Cardiff and were very poor. His mother, Mabel née Geach, had been a brilliant mathematician and scholar of Girton College, Cambridge, and in her final exams was placed first equal with the Senior Wrangler in mathematics at Cambridge, in the days when they did not award degrees to women. She later married a quiet and very devout Welsh Methodist railway clerk, Norman Williams, and the seven girls and two boys (another boy died young) were brought up to be highly self-reliant and all received a good education under Mabel's watchful eye. Dad became a very fine athlete and played rugby for Penarth when sixteen (and six foot) when the Penarth team had eight international caps playing for them. When Norman Williams learned that his youngest son Evan Rhodri was becoming keen on an Irish Catholic girl at Cardiff University, he tried to warn him off this most undesirable relationship: 'She's Irish, you know, and what's worse, she's a Catholic!' Not to be intimidated, dad promptly invited my mother home, where Norman overhead her talking in Welsh to Gladdie. 'Oh! So it's Welsh you're speaking, is it? Come on then, let's hear you!' My mother rose to the challenge, recited one of the Old Testament psalms in Welsh, then a Welsh poem, and even offered to sing a hymn in Welsh. (It was just as well she didn't, since she hadn't a note in her head.) From then on she could do no wrong, having won over Norman entirely.

Mum and dad became engaged some years later, when they both had secure jobs, she teaching and he working as a railway traffic apprentice. During her final Honours year studying French and German, mum's Professor suggested she spend a year at the university in Berlin, to perfect her spoken German. The year was 1932, so she went over as governess to Hilda, the headstrong daughter of Herr and Frau Meissner. During her first day at the university, mum, still unable to read the signs in Gothic scrip (*deutsche Schrift*) on the doors of the lecture rooms, blundered into a physics lecture where a wild looking man was scribbling equations on the board. At the end of the lecture, my mother, as the only woman present, was pushed to the front and dragooned into giving the vote of thanks, having not understood a word of the presentation. 'But who is he?' she asked. '*Das ist Herr Doktor Einstein!*' Mum once again rose bravely to the occasion. In her best German she thanked Einstein very warmly, adding: '*Jetzt ist mir alles klar*!' ('Now everything is crystal clear'). Her expression later became a family saying whenever we found something to be utterly obscure.

In 1932, Otto Meissner had been very contemptuous of Hitler, at least at the dinner table. But he and many other well-educated Germans had seriously under-estimated Hitler, who came to power in January 1933. Many politicians fell under his spell or saw which way the wind was blowing. By December 1937 Meissner had become Minister of State in the German Cabinet. Mum returned once more to Berlin in 1938 at the request of the Meissner's and stayed in their palatial home. At night SS troopers with fixed bayonets mounted guard outside her bedroom door. Realising finally that she was virtually under house arrest, she managed to post a card in Welsh to my father, to whom she was engaged. He was cycling through Europe on a camping trip and was then in Switzerland. Here he received her SOS, caught the train to Berlin, where Meissner lent him a dinner jacket, and over dinner that night dad told bare-faced lies about the highly advanced state of Britain's military preparations for any eventual war. Next morning he and his fiancée, my future mother, left Germany. Unlike mum, who was politically naïve, dad as a well-informed territorial officer and professional railwayman, was convinced that war with Germany was imminent.

September 1, 1939. Following the German invasion of Poland, Britain and France jointly declared war on Germany. My elder brother Tim was born next day. Dad joined the Royal Engineers and was still present at Dunkirk several days after Winston Churchill had announced that the last British soldier had been evacuated. He hitched a lift back to Falmouth in Cornwall on a Grimsby trawler that appeared out of the early morning mist seeking any last survivors. He was ordered to report to the War Office in London. Dazed by many days with very little sleep he asked a London bobby for guidance: 'Which side [of the River Thames] is the War Office on?' 'Ours, I hope, but I sometimes wonder!' was the reply.

Once the German bombing began in earnest, mum led a nomadic existence, from Hertfordshire, where I was born on 19 May 1941, to Barry and finally to Portstewart on the coast of County Londonderry ('Derry') in Northern Ireland (Fig. 1.1), where my younger brother Aidan was born. I spent all my spare time with the gnarled old fishermen, walking hunched up, spitting from time to time, looking up at the sky, and muttering wisely 'Tis so, indeed, 'tis so!' Meanwhile, dad was given the task of arranging the potential invasion of Eire. In the event, an unnecessary task, since his Irish opposite numbers indicated that they much preferred to remain neutral and would certainly not welcome any German invasion. He then served in North Africa, and took part in the invasion of Sicily. When he finally returned to Portstewart he narrowly escaped injury. My younger brother Aidan, ever protective of his mother, hurled half a house-brick at the man in army uniform with his arm around his mother. 'Good God! What was that?' 'Ah, yes! Come and meet your youngest son!'

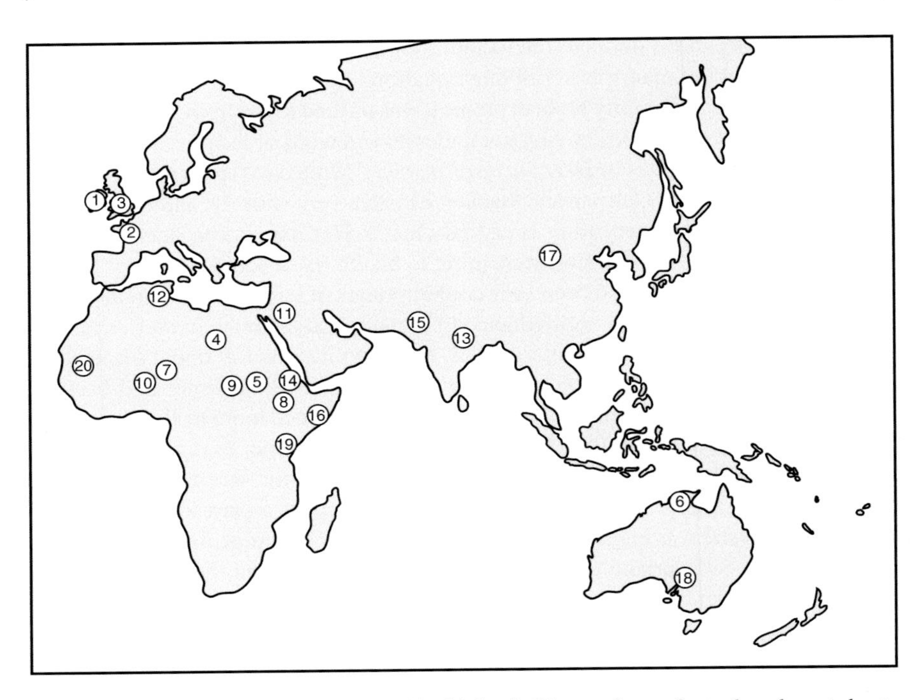

Fig. 1.1 Map showing locations discussed in this book. The *numbers* refer to the relevant chapter

Another shock for dad was finding that all the money he had thought carefully saved in the joint bank account left with my mother while he fought in North Africa and Italy was gone. Mum was as feckless as he was frugal, and from then on there were no more joint accounts! He immediately decided that they both had to work. We left Northern Ireland by ferry from Dun Laoghaire in County Dublin to Fishguard in Wales, where we three boys spent a fine free wild year in a lovely rural boarding school at Davington in the Usk valley of Monmouthshire, South Wales. Mum and dad worked hard in England to save enough for us all to move to a semi-detached council house in Chelmsford, Essex, where we went to a more regimented type of primary school for the next three years. At every opportunity I would play truant and go down to the Baddow Meads along the river. After a visit to the Roman remains at Colchester and after being awarded a Sunday School prize booklet on Ancient Britons and Stone Age tools, I began searching for flints and tried in vain to start a fire by striking blocks of flint together. There were plenty of cuts and sparks, but, fortunately, no fire.

One day in 1950 after he had been away for a short time in France, dad returned home with a French ham and said we were leaving Chelmsford and going to live in France. That ham was the most delicious thing I had ever tasted, so I needed no persuasion to leave. Post-war England was austere, rationing was the rule, we grew our own vegetables on an allotment, and the school lunch potatoes looked and tasted like yellow soap. I was glad to move on!

Chapter 2
Boyhood in France (1950–1953)

Dad had accepted a job in Paris to work for British Railways in France. It sounds so much better in French: Les chemins de fer britanniques en France. In 1950 we moved to a small village called Saint-Leu-la-Forêt, which at that time was set amidst the vast Forest of Montmorency, where the French monarchs used to hunt until July 14, 1789 when the French Revolution and the Fall of the Bastille prison put a stop to all that. I was entranced by the enormous quantities of autumnal leaves to be fashioned into castles and redoubts. I was less entranced when it dawned on me that I was actually expected to learn French. For a while I resisted, and remained at the bottom of my class while my brothers both did well. Mum and dad were both fine linguists, having studied Modern Languages (French and German) at university (both Cardiff and Oxford, in dad's case). They also had a good grasp of Latin and

M. Williams, *Nile Waters, Saharan Sands*, Springer Biographies,
DOI 10.1007/978-3-319-25445-6_2

Fig. 2.1 La-Celle-St-Cloud, France, ca. 1951. I am wearing the grey smock worn by all French primary school boys at that time and have the regulation *cheveux-en-brosse* haircut

Italian, and dad had taught himself Ancient Greek as a boy at Cardiff High School when banished by one of his teachers from history lessons: 'Williams, out!'

Fortunately for me, we moved from Saint-Leu-la-Forêt, where the teacher and I shared a mutual antipathy, to the hamlet of La Châtaigneraie adjoining the small town of Vaucresson near the village of La-Celle-St-Cloud. Dad used to cycle each day to and from the station at Vaucresson and then by train to and from Paris, while we three boys went to the primary school at La-Celle-St-Cloud (Fig. 2.1), where I finally decided to master French and discovered the joys of wide reading in several languages. There was a prison at the end of our street and there were occasional escapes. One day the front gate bell rang while we were at school and dad at work. A very polite escaped prisoner said to my mother: 'Madame, I am taking the small change you left out for the milkman. I felt I should let you know.' Mum thanked him for his thoughtfulness. A little later mum gave some gardening work to an itinerant unemployed man (there was much poverty in post-war France) and he dug up some German helmets from our back garden. The word spread that we were not British after all and that *les Boches* (the Germans) were back. Because mum spoke French with a Welsh and not an English accent, it took some time for these rumours to be allayed. We boys were oblivious to all these undercurrents until mum told us about them much later.

That our appreciation of landscape closely reflects our childhood experiences was brought home to me during a chance visit (during a geology conference in 2008) to the travelling display of the French Impressionists, then on view at the Museum of Fine Arts in Houston, Texas, where I was entranced by Alfred Sisley's *Path through the woods, La-Celle-St-Cloud*. I knew that path, knew it well, for it was uncannily like the one that my brothers and I had followed daily on our long walk to primary school in La-Celle-St-Cloud village some seventy years later. We used to call it 'the school in the woods' and it was here that my superb teacher, Jacques Degorce, drummed into me the names of all the main rivers and the elevations of the highest mountains of France. Oddly enough, and despite this capes and bays approach to geography, he kindled in me a love of landscape that has persisted ever since. Perhaps this was also due to the exploits of the intrepid French mountain climbers Maurice Herzog, Lionel Terray and Gaston Rébuffat on Annapurna at that time. Who knows? The French philosopher Henri Bergson (1859–1941) has expressed what I am fumbling to convey so much better: 'The route we pursue through time is strewn with all we have begun to be and all we might become'.

Thanks to the work ethic and knowledge instilled in me by Jacques Degorce, I won a scholarship to the Lycée Hoche in Versailles. Here I soon realised that the boys who did not work vanished at the end of each term, so there was a powerful incentive to knuckle down and work. The standards were very high and the teachers first rate. One exercise I was set at age eleven was to go to the Louvre in Paris, visit the Egyptian Antiquities rooms in that great museum, and write a ten page illustrated essay summarising what I had seen. Perhaps it was that experience that sparked an early desire to visit Egypt one day and see the temples and frescoes for myself.

While in France we enjoyed the ridiculously long French school holidays. We were expected to work and complete our 'cahiers de vacance' (holiday notebooks), but still had plenty of time for mischief. Our parents were keen travellers and both loved the sea so we took our summer holidays in Spain at Castelldefels and Tossa del Mar on the Costa Brava, at La Napoule near Saint-Raphaël on the Côte d'Azur, and at Quiberon in Britanny. It was in Spain that I learned an important lesson. One day my parents befriended a German couple who were soaking up the sun on the beach near the hotel where we were all staying. Herr Henzehler had been a fighter pilot during the war. I could not understand why my parents were so obviously enjoying the company of "the enemy". Only a few years earlier he and my father had been fighting on opposite sides. Was the war really over that long ago? It was a salutary experience for one small boy.

Another lesson I was to learn a few years later. All the European history I was taught in France was a catalogue of a thousand years of glorious French victories and ignominious English defeats, with the battle of Fontenay taking pride of place in this eulogy to the French fighting spirit. Waterloo was simply not mentioned, nor was Trafalgar. Several years later, after we had moved to Sheffield in South Yorkshire at the end of 1953, I had to re-learn all my European history. Now it was a thousand years of French defeats and English victories. And so, by the age of thirteen, I had acquired an enduring scepticism for anyone (or any institution) claiming a monopoly of truth. This unwillingness to accept authority unquestioningly has stayed with me to this day, and has more than once landed me in hot water!

Chapter 3
Sheffield, the Pennines and Cambridge (1953–1962)

In late 1953 we moved from France to the city of Sheffield in South Yorkshire, renowned for its manufacture of high quality steel and cutlery. There was even a VIP train called *The Master Cutler*. One day the Lord Mayor of Sheffield, a rough, tough, outspoken character, complained bitterly to my father: 'Last week ah took Master Cutler to London; it were three minutes late!' To which my father replied solemnly: 'I'm sorry to hear that, Lord Mayor. I'm afraid I shall have to ask you not to travel on it again!' They parted amicably.

Sheffield in winter we all found pretty grim, with dense smog reminiscent of Dickensian London. Dad soon got us enrolled in King Edward VII School, an excellent grammar school with bright kids and outstanding teachers. I was horrified

M. Williams, *Nile Waters, Saharan Sands*, Springer Biographies,
DOI 10.1007/978-3-319-25445-6_3

on my first day there to see one of the masters caning a pupil with some ferocity; in France there was no corporal punishment. Another difference was the emphasis on school sports and physical education; again, none of that in France. Still, we adapted, although I always felt something of an outsider, and preferred to keep my own counsel than follow the schoolboy party line.

There were three school Scout troops and scouting was something I enjoyed immensely, gaining my Queen's Scout badge and seeing a great deal of new country during wide-ranging camping trips. One such camp was on the headland between Falmouth and Mylor in Cornwall, on land owned by Commander Trefusis, formerly of the Royal Navy. Here we saw the bowling green, one of many, where Drake was alleged to have been playing when his men reported sighting the sails of the invading Spanish Armada in the distance. 'We'll finish the game and then fight the Spaniards!' In the Mylor churchyard there was an old birdbath. Inscribed on it in Latin was the lovely invitation: *Hic licet, O volucres, vos lavare; si solvere vultis, carminibus solvite*! ('Here, little birds, you may wash freely; should you wish to pay, pay with a song'). I may have it wrong, but that is what stuck in my memory. Many years later two of my aunts came to Mylor after they had retired, including Dr Katherine Williams, who had been the Chief Medical Officer at Harwell and who had pioneered safer forms of radiography.

Other camping trips were to Newstead Abbey in Sherwood Forest and to the North Yorkshire moors, not far from Ilkley Moor of the well-known song. The highlight for me was Loch Torridon in Ross and Cromarty, a sea loch in the Highlands of Scotland near Achnasheen, overlooked by the great gaunt mountain masses of Beinn Alligin, and Liathach the grey one, thought by many to be some of Scotland's finest mountains. I learned to fish in the sea loch, and discovered the hard way all about tides when my patrol's chosen campsite became flooded. The local laird guided us to the summit of Beinn Eighe, later to become Britain's first national nature reserve (1951) and my first serious mountain walk. Later, with the senior scouts, we went on a long camping walk around the Ring of Kerry in southwest Ireland, a spot I was to revisit when at university (Fig. 3.1a–c). Like my father before me (he had been a King's Scout and a Rover Scout), I found scouting to be very practical and very useful, especially the advanced First Aid we learned from a professional ambulance officer. I belonged to C Troop, of which the Scoutmaster was Mr O.R. Johnston, a charismatic man with a great love of the Goon Show. He was aided by some enthusiastic and first rate Troop Leaders and Assistant Scout Masters, including Michael Gagan, Trevor Crisp (who taught me to fish for pollack from a canoe in Loch Torridon and became a world authority on salmon) and Tony Guénault, all of whom later enjoyed distinguished scientific careers.

The local people proudly referred to Sheffield as a dirty picture in a beautiful frame. They were right. The bombed out buildings along Froggatt Street in the city centre could hardly be called attractive, but where we lived out on the western margins of the city the moors of the Pennines were only a short walk away. I soon learned to catch the train out to Edale to go tramping on the rugged Millstone Grit country around Kinder Scout. Later I discovered the joys of caving and exploring

Fig. 3.1 a–c Field sketches of the glaciated Macgillycuddy Reeks, County Kerry, south west Ireland (July 1961)

the sites of the old Roman lead mines in the Derbyshire limestone country. In the entrance to Peak Cavern in Castleton, where I stayed in the local Youth Hostel, traditional rope making was still practised, at least for the benefit of visitors. In this cave the Romans had mined Blue John to make vases from this semi-precious form of fluorspar. Out in the Derbyshire plague village of Eyam in 1665 the Reverend William Mompesson had kept the plague from spreading through the Derbyshire dales by enforcing a strict quarantine. History was everywhere in these dales. With my fellow scouts or on my own I hiked through the hills and dales of Derbyshire, mostly camping, and learning to travel light. Although I have dabbled with caving, rock climbing and snow and ice mountaineering, walking has remained my favourite form of outdoor activity ever since my boyhood days in the Pennines amidst wind and rain.

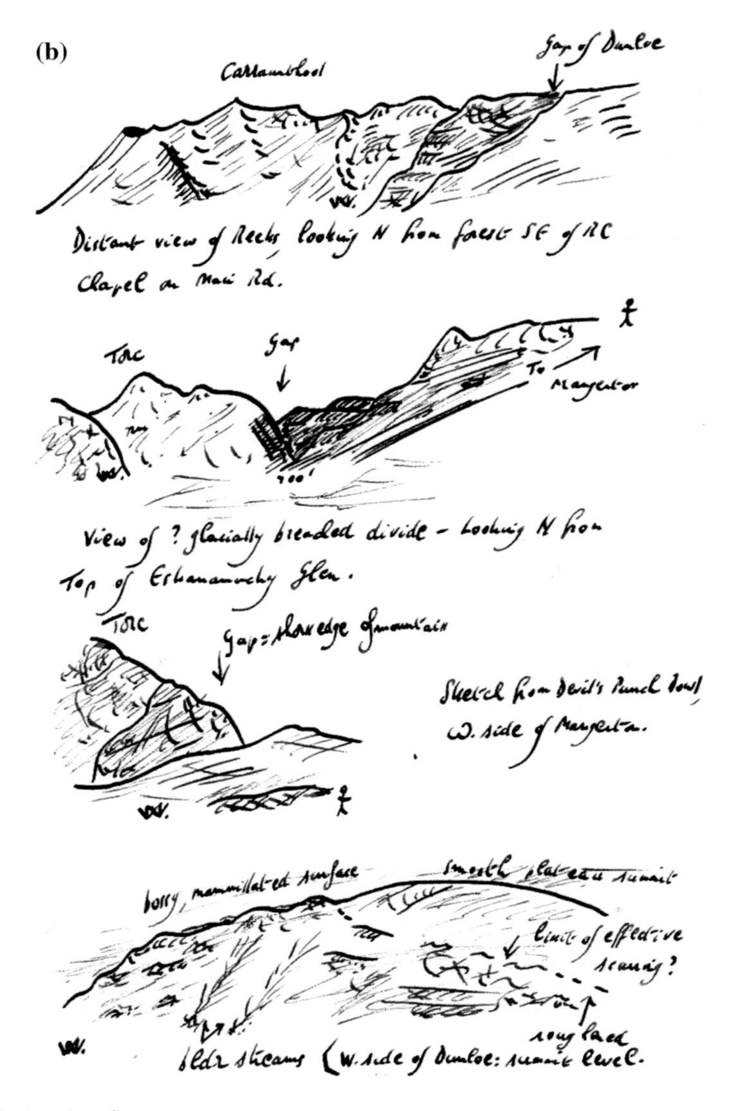

Fig. 3.1 (continued)

I enjoyed my time at school and benefited from superb teaching by dedicated, enthusiastic and very bright teachers. There was no subject I did not enjoy, even maths, where my teacher Geoff Ingham was twelfth man for Yorkshire and as adept at throwing chalk as he was as an all round cricketer. Bert Towers taught us regional geography in his gently lugubrious way, as well as how to survey and make maps. Dennis Henry, a brilliant Irish classicist, was outraged if we did not score close to 100 % for Latin. Bert Harrison encouraged me to compete for the

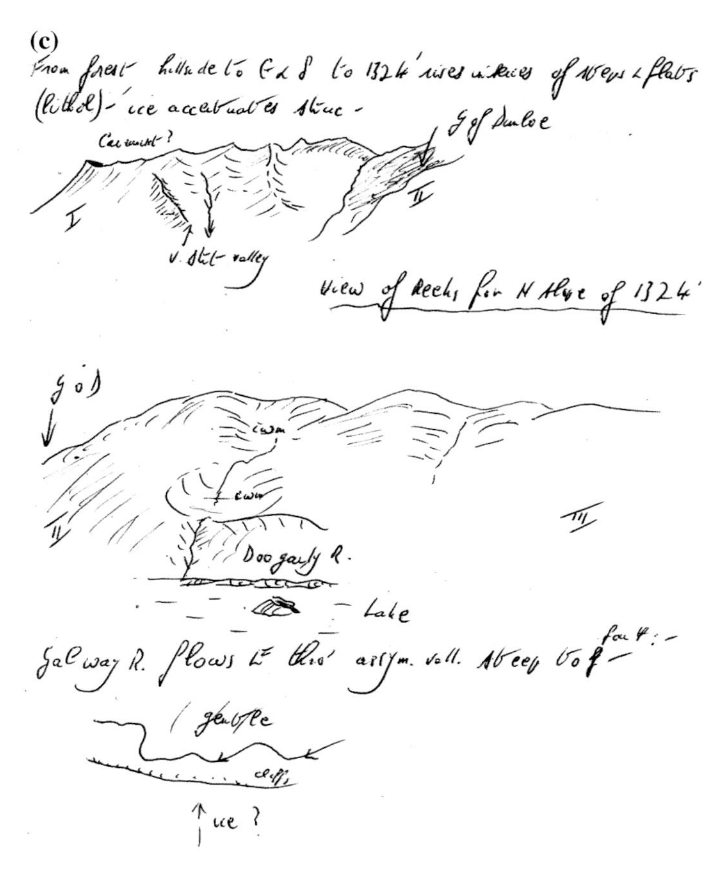

Fig. 3.1 (continued)

bronze medallion for life saving, and taught me the rudiments of judo, which I later pursued at university. I found it hard to decide whether to go for science or arts, and finally opted for geography, winning an open exhibition in geography to Selwyn College, Cambridge in 1959.

I had a happy three years as an undergraduate student at Cambridge and discovered the joys of wide reading and endless talking! The Master of Selwyn College in my time was the Reverend Professor Owen Chadwick, a brilliant history scholar and the former captain of the Cambridge rugby team. The distinguished engineer Sir David Harrison succeeded Owen Chadwick as Master of Selwyn; he had been my 'tutor' in my first year in college and was a gentle and profoundly civilised man.

Bruce Sparks was my academic supervisor. Once a week I read out the essay I had written that week while Bruce peered down his microscope at the non-marine fossil snail shells he had collected from across East Anglia. In my final year my friend Nick Large, a scholar from St Johns, shared supervision with Bruce. Nick and I used to cycle out to sand and gravel quarries in search of fossil sharks' teeth. On summer afternoons I would play a friendly game of tennis with Roger Bristow, a fellow geographer with whom I shared workspace in my final year. I was fortunate to be taught judo by Kenshiro Abbe, a former world champion; from him I learned the many ways to fall without injuring myself—a skill that probably saved me from serious hurt on two occasions later on.

The lectures I most enjoyed were those that dealt with the physical side of the discipline, quaintly called 'physiography' in those days. Some of our lecturers had been involved in intelligence work during WW2. My Selwyn College tutor in my second year at university, Dr. J.K.S. St Joseph had been an ace at air photo interpretation. After the war he pioneered the use of air photos to interpret archaeological sites in Britain and Ireland. W.W. ('Bill') Williams had applied his knowledge of coastal processes to advising on the Normandy landings in 1944. Bruce Sparks had served in the Navy and taught us meteorology. However, my fondest memory is of mapping glacier flow in the Val d'Hérens in Canton Valais in southern Switzerland in the summer vacation at the end of my second undergraduate year. Jean and Dick Grove ran this wonderful fieldtrip. Jean was a glacier specialist while Dick enjoyed surveying. He had worked as a soil surveyor in northern Nigeria and sparked my interest in the landforms of the Sahara and its borders. Dick would invite guest lecturers to his course, one of whom, his Ph.D. student Claudio Vita-Finzi, gave a vivid account of the old Roman ruins of Tunisia and Tripolitania in northern Libya—ruins I was to see for myself a few short years later.

Alfred Steers was Professor of Geography and in his lectures took us across every inch of the coast of England and Wales. On one occasion he saved my life by whisking me from the path of an oncoming lorry as we were walking along a narrow lane to the geography department in Downing Place. His successor was the effervescent Dick Chorley, whose scurrilous accounts of the great pioneers in the discipline of geomorphology had us in stitches. More seriously, he taught us basic statistics, including regression analysis from first principles. He was later to spearhead the revolution in quantitative geomorphology based on some of the laws of elementary physics, to the delight of some and the derision of others. At all events, he kept us interested, entertained and stimulated. What more can you ask from a teacher?

We had excellent and enthusiastic lecturers in the urban, historical, economic and regional aspects of geography (Gus Caesar, Christopher Board, Benny Farmer, Peter Haggett, Jean Mitchell, Clifford Smith, Harriet Steers, Tony Wrigley), but somehow these branches of the discipline never exerted the same attraction for me as the study of landforms and landscapes.

There was a strong interest across a number of departments (geography, geology, botany, archaeology, zoology) in geologically recent climatic fluctuations, and a series of lectures were given on these topics. In both my second and third year I was enthralled to hear Harry Godwin and Richard West talk about reconstructing vegetation and climatic history from analysis of fossil pollen grains, Richard Hey talk about his geo-archaeological work at Haua Fteah cave in northern Libya, and Bruce Sparks discuss his studies of fossil mollusca. After attending these lectures and demonstrations, I became more and more interested in recent geological events.

The PhD students working on ice, dunes, erosion rates and limestone caves were great role models for a keen young undergraduate. They included Paul Williams, who became a world authority on limestone caves, whom I met later during his time at the Australian National University and when he was Professor of Geography at Auckland University in New Zealand; Mike Kirkby, an energetic fell walker from the Lake District and a brilliant mathematical modeller, who later edited the journal *Earth Surface Processes and Landforms* with great distinction for many years; Olav Slaymaker, who went on to great things in Canada. Another PhD student, Andrew Warren, was studying windblown dust deposits in East Anglia supervised by Dick Grove but decided that the dunes of Kordofan Province in Sudan were more interesting. He encouraged me to apply for a job with Hunting Technical Services Limited, having himself worked for them investigating saline soils in the Indus valley of Pakistan. It is thanks to Andrew that I found myself mapping soils in the Sudan shortly after graduating, as I shall describe later (see Chap. 5).

Chapter 4
Expeditions to the Libyan Desert (1962–1963)

One of my most vivid memories of growing up in France was the sheer abundance of high quality popular articles relating to exploration in North Africa and other parts of the French-speaking world of the day. One such article, which had an abiding influence upon me, was by the great French Saharan geographer E.F. Gautier, with the alluring title of 'the dead rivers of the Sahara'. (The French took their *haute vulgarisation* very seriously in those days, and happily still do.) Reading Gautier's account succeeded in convincing one small boy that he, too, one day, would visit these mysterious rivers, which he duly did, in the northern summer of 1962.

The opportunity arose quite suddenly, soon after I had graduated from university, with a telephone call from Dick Grove to my mother in Sheffield (I was caving in Derbyshire at the time) asking if I wanted to join a British Army expedition to Jebel

M. Williams, *Nile Waters, Saharan Sands*, Springer Biographies,
DOI 10.1007/978-3-319-25445-6_4

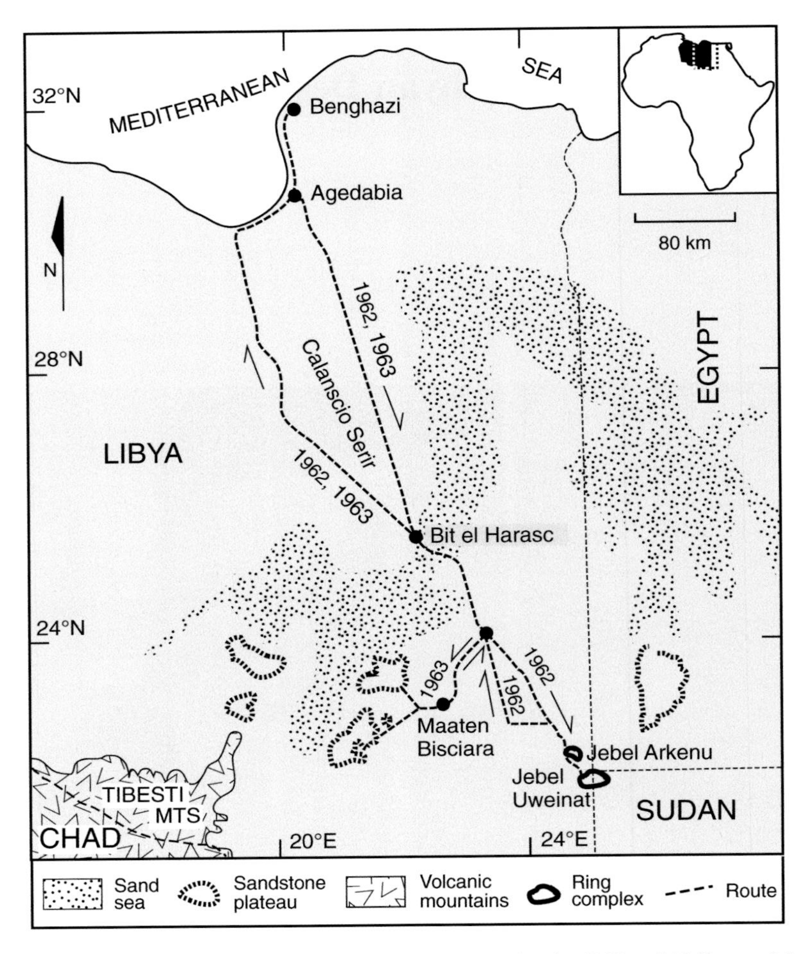

Fig. 4.1 Libyan Desert. *Dashed lines* show routes followed by the 1962 and 1963 expeditions to southeast Libya

Arkenu, a remote mountain in the desert of southeast Libya (Fig. 4.1). Dick had lectured to me as an undergraduate at Cambridge, and after an inspiring field trip with Dick and Jean and their young family, mapping glacier flow at the head of the Val d'Hérens in southern Switzerland, and an earlier walk across the spine of Norway with Norwegian friends, I had become enamoured of glacial landscapes. While at school in Sheffield I had read accounts of past glaciations in Britain and Europe, had walked the beautiful Ring of Kerry in southwest Ireland, and earlier still had rowed with my parents across one of the glacial lakes of Austria. My Honours dissertation (the Cambridge Geography 'regional essay') was on the glacial breaches and sub-glacial channels of the Macgillycuddy Reeks in County Kerry.

So persistently did it rain and so dense was the mist during the fieldwork in Kerry in July 1961 that I seldom caught a glimpse of any of the glaciated high peaks. (I was able to remedy this omission during some rare cloudless days, many years later.) The experience of what the Irish euphemistically call 'beautiful soft weather' predisposed me favourably towards arid regions where there seemed to be some guarantee of keeping reasonably dry, so I was happy to offer an immediate and enthusiastic yes to Dick's invitation. (Much later, the heavy rains I experienced in Inner Mongolia in August 1999 and the equally heavy rains in Tunisia and Algeria in January 1970 and in the Mauritanian desert in January 2004 were to command my unreserved respect for the power of torrential rain and flash floods in deserts.)

Captain (later Lieutenant-Colonel) David Hall RE (Royal Engineers) was in command of the Sandhurst expedition to Jebel Arkenu in 1962. David was an instructor at the Royal Military Academy, Sandhurst at Camberley in England and a great desert enthusiast. We soon became firm and lifelong friends and I owe a great deal to David for introducing me to the Sahara. Having driven down through France and eastwards along the Libyan coast from Tripoli to Benghazi (Fig. 4.1), we set forth from Agedabia across the flat and featureless gravel plain of the Calanscio serir to the tiny grove of palm trees at Bir el Harasc Bir (Fig. 4.2a–d). Here we obtained fresh water at shallow depth, replenished our water jerry cans, and set forth for Kufra oasis, where the main road was built of rock salt (showing how seldom it rained) and where we enjoyed a blissful swim in one of the lakes fed from hot springs. Numerous small boys at Kufra brought us scorpions in various states of disrepair for our London Natural History Museum collection and while their elders had our attention the smaller fry cleverly relieved us of some of our possessions. We never realised how cunningly we had been distracted and duped by these light-fingered urchins until our return to Kufra some weeks later, when the police chief offered us glasses of sweet, mint-flavoured tea. With a glint in his eye, he opened a cupboard and asked if we recognised anything. There, carefully arranged, were all our missing possessions!

I had earlier asked Dick Grove what I needed to do on reaching Jebel Arkenu. His reply was commendably brief: 'Everything!' So I came prepared to map the mountain as best I could using whatever air photographs were available. I also planned to collect rock samples to help prepare a rough geological map of this intriguing mountain (known to geologists as a ring complex), and to record any evidence of prehistoric human occupation in the form of rock art, stone tools and other archaeological remains. A small team of fit and highly competent officer-cadets and specialist and highly professional soldiers helped us greatly in achieving these aims and in keeping the vehicles and radio in good running order, while our effervescent medical doctor Peter Beighton (later Professor of Genetics at the University of Cape Town) kept an alert eye on any health needs, most notably a few cases of heat exhaustion and dehydration. Shade temperatures were in the mid to high forties and we were seldom in the shade. Water consumption per person was about ten litres (two gallons) a day. Washing became an optional luxury but in the dry heat never seemed very necessary. We slept on the ground and apart from an

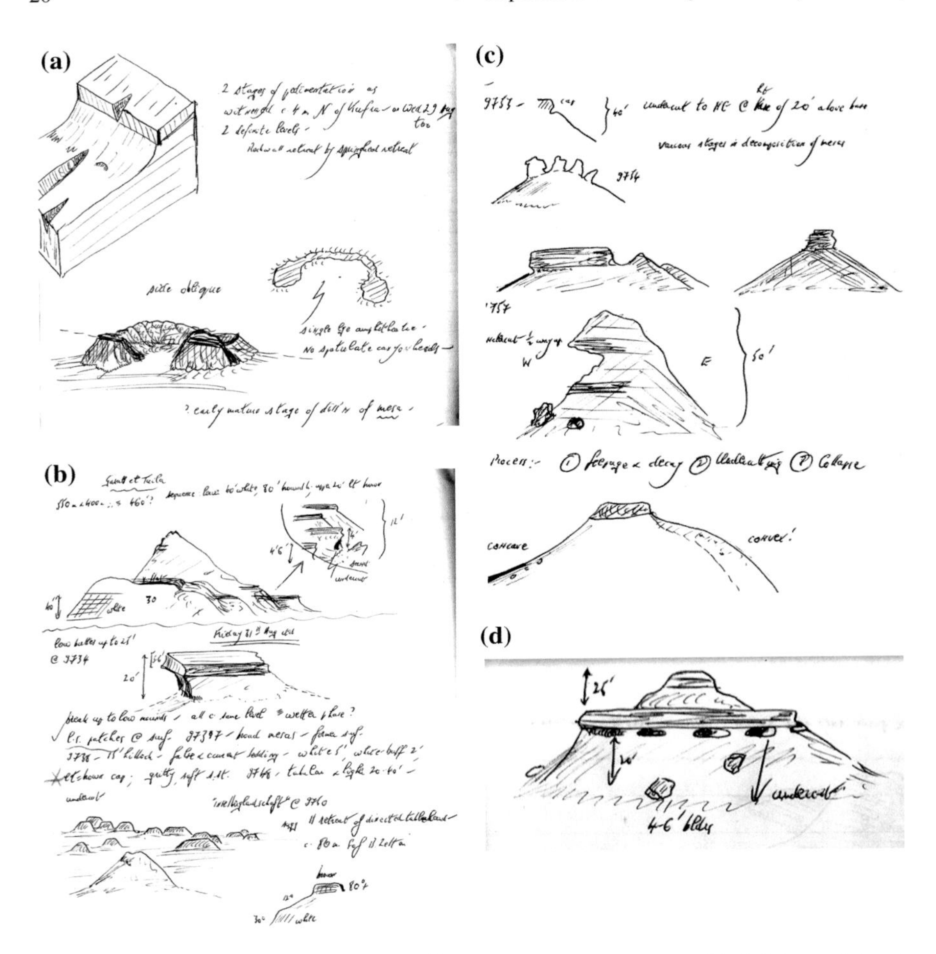

Fig. 4.2 a–d Field sketches of desert landforms seen when en route to Jebel Arkenu, Libya in 1962

occasional nocturnal visit from an inquisitive scorpion we suffered no discomfort. Others had visited the Jebel before us: in 1923 Hassanein Bey found fresh water from a spring high up the main wadi, but by 1962 the former spring had dried out leaving an unappealing residue of rock salt and gazelle dung. Fortunately, we found abundant good water from an underground source at Jebel 'Uweinat located 40 km to the southeast astride the borders of Libya, Egypt and Sudan.

A year later we returned to southeast Libya, this time to explore and map two large sandstone plateaux (Fig. 4.3a, b) apparent on air photos flown by the RAF during the Second World War (WW2) but not featured on any of our small scale WW2 maps, remarkable chiefly for their frequent mention of 'relief data incomplete': a magnificent understatement. Using a reasonably complete set of air photos

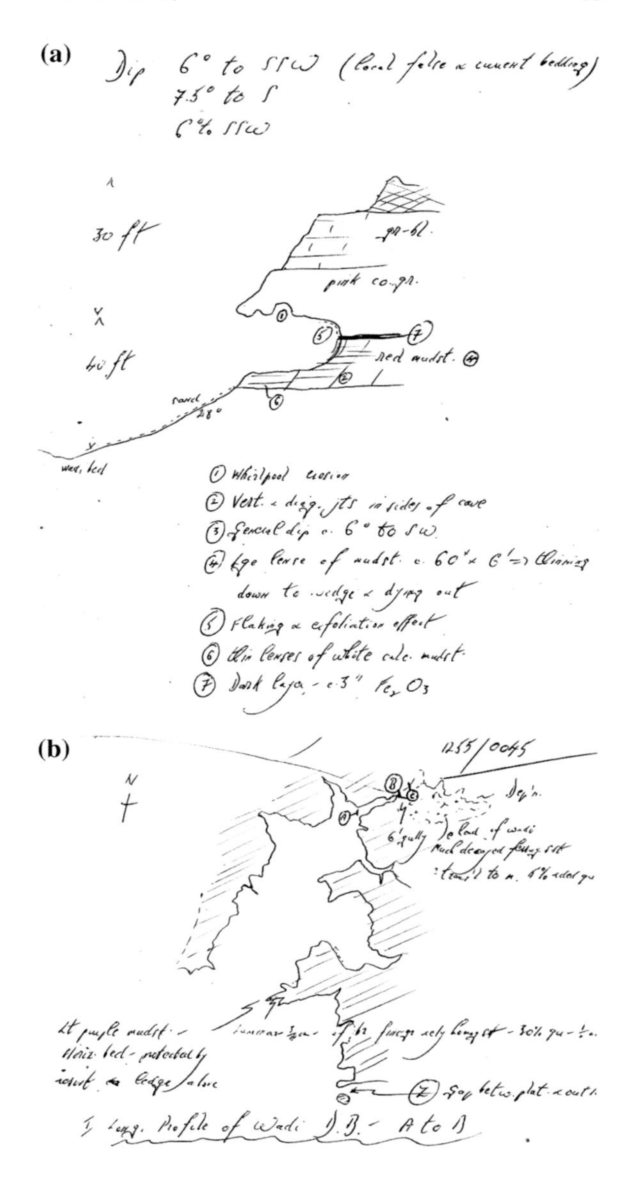

Fig. 4.3 **a**, **b** Field sketches of Nubian Sandstone landforms mapped during 1963 expedition to southeast Libya

spread out on the floor of a metal Nissan hut (masquerading as an oven) in the Benghazi barracks, I prepared a rough outline map of the two plateaux. We set forth along our usual route as far as Kufra, whence we drove southwest to the deep well at Maaten Bisciara and so on to the two great sandstone tablelands, which covered an area of about 200 by 80 km, separated by a valley up to 20 km wide. David Hall was again in command, with John Blashford-Snell RE as his second in command

Fig. 4.4 David Hall (*left*) and John Blashford-Snell, southeast Libyan Desert, August 1963

(Fig. 4.4). John was a lively character and a British Army pistol champion with a disconcerting habit of firing his pistol at any small desert mammal he might see in our vicinity. A tiny jerboa nibbled his ear one moonlit night when he was fast asleep, which caused him to charge off starkers into the desert firing his pistol and uttering troopers' oaths, providing us some mischievous delight despite our own broken sleep.

We gave Arabic names to the two plateaux and some years later the Italian geologist Angelo Pesce identified a third sandstone plateau situated further south, which he named in honour of the great Arab traveller Ibn Battutah (1304–1377) whom I and many others consider the greatest desert traveller of all time. Pesce worked for the Petroleum Exploration Society of Libya and had access to the stunning Gemini Space Photographs of Libya and Tibesti, which ushered in a revolution in desert exploration and mapping. Satellites eventually meant universal access to Global Positioning System (GPS) hand-held devices, which are now routinely used for navigation. In the early 1960s our desert navigation relied on dead reckoning using the Bagnold sun-compass by day, with star fixes by theodolite at night to confirm location, just as the Long Range Desert Group New Zealanders and British had done during their forays deep into the Sahara and behind enemy lines during WW2. Seeking a few palm trees on a bearing of 172° over a distance of 250 km needed considerable skill, which David possessed to a high degree. The palm trees meant water at shallow depth. If we missed, we were in trouble. Without water one does not survive long in the middle of the Sahara in summer.

Fig. 4.5 Mud curls on surface of former small lake impounded between dunes, southeast Libyan Desert, August 1963

Once again, as at Jebel Arkenu the year before, the most striking thing we discovered about the two sandstone plateaux was the abundant and widespread evidence that prehistoric humans had occupied these now bleak and barren tablelands. We found Neolithic pottery, grindstones, stone circles and ostrich eggshell beads, as well as what appeared to be old water channels or possible irrigation ditches on the surface of the plateau and ponds that once existed between dune barriers (Fig. 4.5). In fact, as I was to appreciate later, scattered across the Sahara from the Atlantic coast of Mauritania to the Red Sea Hills of Egypt and Sudan—a distance from west to east of 5000 km—there are sporadic stone tools that range in age from Early Stone Age (Lower Palaeolithic) through Middle Stone Age (Middle Palaeolithic) and Late Stone Age (Upper Palaeolithic) to Neolithic, pre-Islamic and historic. The Saharan Neolithic sites are on occasion associated with rock paintings showing herds of brindled cattle, dogs, men armed with bows and arrows, cattle camps, women riding oxen, even canoes apparently made of papyrus, all of which suggest far wetter climatic conditions at the time of pastoral activity in what is now a hyper-arid wilderness. Indeed, the Arabic word *sahra* means a deserted wilderness to be traversed as rapidly as possible.

And so, by the end of 1963, I had begun to ask myself why the great Sahara desert—the largest hot desert on earth—had once been home to groups of Neolithic nomadic pastoralists with their herds of cattle and their hunting dogs, and why this was no longer possible. Had the herdsmen brought about their own demise through overgrazing and accelerated loss of plants and soil? Or was the cause more fundamental, and linked to long-term changes in climate? If so, what were the causes? And what was happening in other parts of the world such as Australia, China and India? Had the deserts in those regions previously been more hospitable also?

Chapter 5
Blue and White Nile Valleys, Sudan (1962–1964)

Soon after returning from the 1962 Sandhurst expedition to Jebel Arkenu in southeast Libya, I had accepted a job with the British company Hunting Technical Services Limited to work as a soil surveyor in the Blue Nile valley of the Sudan. I received my written instructions and documents at Heathrow Airport, where I met Laurie Henderson, a Scottish geologist and one of the team, and flew to Khartoum. Within a remarkably short time the soil survey team of Laurie, myself, Roy Law and Eric Lawrence (both seconded from the Macaulay Soils Institute in Scotland) had begun work under the experienced eye of Colin Mitchell, a delightful man, who spoke fluent colloquial Sudanese Arabic, read the Old Testament in Arabic, and had worked many years in Sudan. Colin later became a senior lecturer in soil science at the University of Reading.

M. Williams, *Nile Waters, Saharan Sands*, Springer Biographies,
DOI 10.1007/978-3-319-25445-6_5

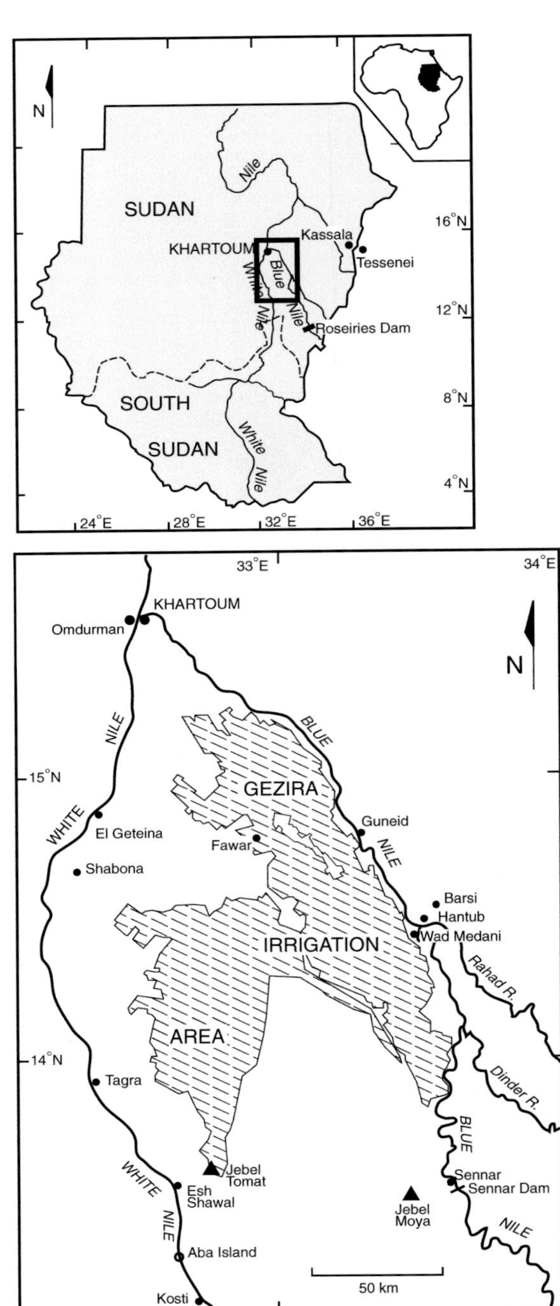

Fig. 5.1 Blue and White Nile valleys, Sudan (1962–1964)

Fig. 5.2 Canal junction, Gezira Irrigation Area, Sudan. *Photo* Don Adamson

Fig. 5.3 Major canal, Gezira Irrigation Area, Sudan. *Photo* Don Adamson

We were based at first in a dilapidated rest house in the village of Fawar, set amidst the canals of the northern Gezira Irrigation Area (Fig. 5.1). Gezira is Arabic for island and refers to the land between the Blue and White Nile rivers, bounded to the south by the Sennar-Kosti railway line. The Gezira irrigation scheme (Figs. 5.2 and 5.3) began as a series of cautious pilot projects from 1911 onwards but only really took off in 1925 fed by water impounded by the Blue Nile Sennar dam, which was completed on time in July 1925. The entire system was based on gravity flow, and from its inception involved a close partnership between local tenant farmers, government, and technical advisers. It is one of the largest and most successful irrigation projects on earth. With the planned completion in 1965 of another dam on the Blue Nile, this time at Roseires located 280 km upstream of Sennar, and close to the border with Ethiopia, our task from 1962 onwards was to seek out suitable land for future agricultural development to enable the irrigated areas to be extended wisely.

I had only been in the country a week when I experienced my first bit of excitement. One evening, clad only in an old and baggy pair of shorts, I was carefully removing a mosquito from one eye while staring into a steel mirror with a Tilley lamp nearby for light. Glancing down I spotted what seemed to be a stick on the floor near my bare feet and thought no more of it. Having removed the offending mosquito, I looked down again and observed that the stick was in fact a snake and that it was coiled around my left leg and moving slowly upwards. I immediately jumped, kicked out and yelled. Eric came rushing over, saw what had happened, and asked in his lilting Welsh accent: 'did 'e bite yer, then?' to which I replied 'No'. I had felt nothing. 'Oh' said Eric, after a quick inspection, 'he did bite yer!' It was on the sole of my left foot, providentially hardened after many weeks walking in the Libyan Desert. My companions despatched the snake (a poisonous viper) and set forth to find a doctor.

I first applied an old judo belt as a tourniquet cum compression bandage around the site of my femoral artery, slackened and re-tightened about every fifteen minutes. Inspection of the foot showed the two fang marks, about a cubic centimetre (1 cc) of translucent pale yellow fluid on the surface (which I wiped off), and the teeth marks of the viper forming a neatly curved small triangle. (Snakes disarticulate their jaws when they strike). With a scalpel I enlarged the two fang punctures, applied Condy's crystals (potassium permanganate) to the two small wounds to oxidise any remaining poison, asked our cook, Salah, for a cup of tea, and lay down to read a book. About an hour later my companions returned with a worried Sudanese doctor, who gave me an injection of 4 ccs of viper serum and 6 ccs of cobra serum, 'just to be on the safe side'. Until then, I had felt fine! Next morning the villagers came to see my corpse removed. They were duly enlisted to place all our soil survey equipment above ground level and to block any holes that might allow snakes to enter. The local names for various species of snakes left little to the imagination: 'Father of ten minutes' was one, 'shroud-bearer' another.

Soon after this incident an elderly man came to see me complaining of headaches. I asked whether they were worse in the dry season or during the rains, and worse in the morning or the afternoon. His replies suggested sinusitis caused by dust. I pointed to some eucalyptus trees growing alongside a canal, and proposed that he ask his wife to boil the leaves while he then inhaled the vapour, with care to avoid scalding, using his turban to enclose the steam, and gave him some codeine tablets for good measure. It worked, but from then on after work I had a steady stream of patients, having acquired an entirely spurious reputation as a *hakim kebir* or mighty doctor. I already had a reasonable grasp of First Aid but remained petrified that someday I would be confronted with an impossible case.

During our first few weeks in the northern Gezira, Colin very sensibly got us to describe a series of soil pits dug quite close to one another in order to hone our

Fig. 5.4 Sand dune immediately east of the lower Blue Nile, central Sudan

skills and, equally important, in order to check how much the soils varied across the landscape. We had been told, very firmly, by some of the soil survey staff based at the Gezira Research Station in Medani on the Blue Nile that the Gezira was a uniform clay plain, so that all we needed to do was to map and describe soils using a square grid superimposed on air photo mosaics to ensure a consistent mapping density. It all seemed too good to be true, and, of course, it was not. In the event, the northern Gezira proved to be a complex of ancient river channels, sand dunes (Fig. 5.4), gravel deposits and, yes, clay plains. This meant that we had to be extra vigilant in our mapping and choice of sites selected for our soil pits and soil auger boreholes. After we had moved east of the Blue Nile, using as our base another rest house at Hantub, just across the river from Medani, we found ourselves in the vast clay plains traversed by the Rahad and Dinder rivers, two Ethiopian tributaries of the Blue Nile, which shrink in the dry season to a few shallow ponds.

Although describing and sampling one almost identical cracking clay soil profile after another was pretty monotonous work and mentally not very demanding, we had our lighter moments and even occasional excitement. With my fellow soil surveyors on local leave one week, I found myself in charge of our full team of pit-digging labourers. They were a tough and cheerful bunch and came from the two villages of Hantub and Barsi. We began work at sunrise and finished about mid-afternoon. One day out on the Dinder plains, a local village headman (*sheikh*) invited us to come and drink water, by which he meant *merissa* beer brewed from sorghum: very refreshing in hot weather and a bit like weak and slightly sour alcoholic porridge in taste and consistence. A forty-four gallon drum of beer was rolled out and calabashes of beer handed to our thirsty men. The only snag was that the men from one village were all teetotallers, while the men from the other village felt that they had to make up for this deficiency or the sheikh would lose face. One of our drivers dropped his transistor radio into the barrel of beer at the precise moment when an elderly gentleman hobbled in search of free drinks. From the barrel came the plaintive wailing of Radio Omdurman. The old man cast one horrified glance at the barrel, turned and fled. However, we all had a good time, thanked the sheikh for his hospitality and set forth in single-file along a narrow path through some dense acacia thorn forest to reach our vehicles. Suddenly everyone halted. A large cobra had reared up in front of us, hood extended, and in no mood to let us pass. Whereupon one of our men, El Amin, stepped forward, holding a tiny axe, passed the cobra, climbed up into a tree, cut a small branch, trimmed it, walked calmly up to the puzzled snake, tapped it gently on the nose, and told it to go. This the snake did and we set forth once more.

Another occasion nearly proved fatal. We had engaged a new driver in Khartoum and three of us (the driver, our camp clerk and I) were in our Land Rover driving south along one of the many corrugated dirt tracks east of the Blue Nile, heading for Hantub. We never reached Hantub. At one bend in the track the driver overshot, wrenched the steering wheel round and hit the brake pedal hard, a process he repeated. The vehicle rolled over three times and landed upright. I was projected through the left door, rolled across the abrasive sand on the surface of the hard clay beneath, and came to a halt with blood spurting from the artery on my right wrist. I immediately applied pressure to the brachial artery to stop the bleeding, plastered the wound with soil bags, and went to help the dazed driver and camp clerk Mark from the rather battered Land Rover. The ground was strewn with the picks and shovels we had been carrying. Mark, a Christian from the South, muttered that the Trinity had saved us during our three-fold roll. The sunroof helped, too. I was very annoyed at the driver's carelessness, walked to a nearby village, got help to reload the vehicle and drove to the Guneid Sugar Plantation. My bloodstained appearance startled the European occupants of the mess at their afternoon tea. I was offered a cup of tea by a kindly Dutch lady and her husband, a whisky by a forthright Belgian, while the Englishman removed his pipe from his mouth and said: 'I say, old chap, would you like a wash?' I agreed to all three, in that order, as well as to an anti-tetanus jab. I telephoned Pat Wardle, our Expedition Manager in Khartoum, suggesting that I might need a new Land Rover. Next day I drove the battered vehicle back to Khartoum. We later learned that the hapless driver had been fired from the Sudanese Railways for crashing a train: it transpired he could only see out of one eye.

My next driver was Mohamed Bakheit, who had served with the Shendi Camel Corps. He was a very cautious driver and a keen footballer, having played for Sudan. He was a friend of Hashim Dafalla, headmaster of Hantub School, one of the best schools in Sudan. Hashim Dafalla had played for Leicester City in his younger days and was always very supportive of our work. I had some local leave due towards Christmas so Mohamed and I drove to Kassala, a beautiful little town nestling at the foot of the spectacular granite hills of Jebel Kassala (Fig. 5.5a–d), located just west of the former Ethiopian (and present Eritrean) border. While watching the football team from nearby Roma play Kassala, I was asked as a matter of urgency if I could transport a wounded Ethiopian policeman across the border to hospital in the small town of Tessenei. He had been shot in the hand during a border incident up in the mountains involving local bandits (*shifta*). In Tessenei we stayed

Fig. 5.5 **a–d** Jebel Kassala, eastern Sudan

at a small inn run by an elderly Greek who wanted me to play billiards with him morning, noon and night. To escape ordeal by billiards I went for a walk in the mountains. On my return the Tessenei police chief asked me politely if I wanted to be shot. I asked why and by whom. 'Shifta', he replied. 'You are foreign and so you have money'. Compelling logic. In the circumstances it seemed best to cut short our stay, leave the delights of Tessenei, return to Kassala and then on to Khartoum.

The 1962–63 season mapping soils along the former flood plains of the Dinder and Rahad rivers were a useful induction into soil survey procedures as well as introducing me to the language and customs of the peasant farmers and nomadic camel herders of the central Sudan. Much as I enjoyed the life style, I did not find semi-detailed soil surveys very challenging and applied to become a reconnaissance soil surveyor. Somewhat to my surprise HTS agreed and I was allotted a sizeable chunk of land east of the lower White Nile to survey during the 1963–1964 eight-month field season. (The area to be surveyed amounted to 1.31 million feddans, equivalent to 1.36 million acres or 5500 km^2.)

The weeks before the rains could be trying for those in the field. Temperatures were in the forties by day, and the humidity increased progressively. Scorpions took refuge inside my desert boots by night so that early next morning I had to make sure to shake them out. When the rains came in June the clay soils would become too sticky for easy vehicle movement, so this was when the HTS soil surveyors migrated, like swallows in summer, back to our UK base in Boreham Wood, Hertforshire, to write our reports and prepare our albums showing soils, land capability and proposed new canal networks.

The Soil Survey Manual of the U.S. Department of Agriculture was my indispensable guide in those days, and I cannot refrain from quoting one delightful passage from pp. 437–438 of that inestimable volume: 'Excellent soil scientists for some kinds of research, including detailed soil surveys, fail utterly in reconnaissance soil mapping. They may be unable to visualize large and complex patterns and become mentally harassed by indecision in the face of vague and apparently conflicting evidence.'

I loved the life of a reconnaissance soil surveyor and was as happy as a sand boy doing a job that suited my skills and temperament. Moving camp every few weeks with what became a team of skilled and trustworthy diggers, I soon developed a great affection for the people in those parts. The landscape of the Gezira certainly cannot compare in sheer stark beauty with the highlands of Ethiopia and South Africa, or the snow-capped ranges of northwest China, the Indian Himalayas and the Patagonian Andes. True enough, but this semi-arid region has a charm all of its own. My favourite times were around dawn, when a cool breeze would blow, urging us to be up, and in the late afternoon camped in the welcome shade of the tall acacias, with doves cooing, flocks of sand grouse returning to their nests, and stately files of guinea fowl strutting among the trees. Another vivid memory was being out in the tall grass plains east of the Blue Nile during the heat of the day, watching vast herds of camels grazing on the succulent *nal* grass (*Cymbopogon nervatus*) with the distant granite hills of Jebel Fau shimmering in the heat haze.

The local people were well aware of my presence. One day I was driving through the village of El Geteina and stopped for a glass of tea. The tea shelter had a straw roof and was in deep shadow as I entered and greeted the occupants. From near the back an elderly man spoke out. 'You are the *hawaja* (foreigner) who in February [I forget which month] last year put down eight boreholes near [the village of] Hureidana Hag el Musa, and the next day you put down six holes near Tayiba.' I was staggered. 'How did you know? Were you there at the time?' To which he replied, simply and sadly: 'No. I have been here all my life.' I then saw that he was blind.

There were times when I was the unwitting cause of consternation among the local people. One cold and windy morning I was writing in my field note book while sitting on a great heap of clods of clay piled up next to a deep pit dug the previous day which I still needed to describe and sample. I noticed that none of my men would come near me. I asked why. 'You are sitting on a snake', they replied. Glancing down I observed the skin of a snake stuck in the cracks between the clods of soil. 'Ah, yes, blown here by the wind' I commented gaily, and continued writing. They remained adamant that there was a snake beneath me. To allay their fears I took a soil auger and moved aside the clods of clay. They were right. A beautiful saw-scaled viper (*Echis carinatus*) was coiled up, having just sloughed off its skin, until I rudely disturbed its repose. I picked up the snake gently and placed it in a large cotton soil bag, tied the neck, and asked one of the men to tie it carefully under the canopy near the back of the Land Rover. My aim was to send it to the zoo in Khartoum. I had by now become keenly interested in learning the local names, medicinal properties and other uses of all the plants I was collecting.

Wishing to discover which of the plants growing in this area camels found the most and the least palatable, I offered a lift in the rear of the vehicle to an elderly nomad. We set forth. A sudden frantic banging on the sliding window between front and rear alerted me that something was amiss. '*El kis, el kis*' ('the bag, the bag') he yelled, 'it is moving and alive'. To my horror I realised that the bag had not been tied under the canopy, as I had assumed, and that he was sitting on it. Not wishing to alarm him further, I gently advised him that the bag contained a snake. He screamed and leapt out of the Land Rover, and soon disappeared from sight among the thorn trees. He was unharmed but the word spread rapidly that I carried live snakes in my vehicle and for a while the local people gave us a wide berth.

In November 1963, I was based for a few weeks in an ancient rest house on the western flank of the long sand ridge upon which the village of Hashaba was located. The *Umda* of Hashaba was the young and dynamic Idris Habbani, popularly known as the 'Transistor Mayor'. He was responsible for administering a group of villages in the general vicinity of Hashaba and for presiding over the local law court, which he did with humour and good sense. He and his younger brother Khalid soon became my friends and used to visit me every few days in the early evening to see that all was well, bringing me a copy of Churchill's *River War* to further my education. One day Idris, Khalid, their portly accountant and their slim and charismatic *feki* or holy man (who served as their family chaplain) came to me in tears to tell me of the assassination of the youthful President John F. Kennedy on the 22nd of November 1963. I was much moved by their sorrow and concern as well as by the tragic loss of a young life.

Eleven years later, on Christmas Day 1974, I had a somewhat similar experience. Mike Talbot (a geologist friend who was then living in Ghana) and I were camping near a group of Tuareg in Wadi Azaouak in Niger, after having completed some strenuous weeks of fieldwork. It was early morning and several Tuareg came over to invite us to join them for an early meal of camel's milk and millet porridge, washed down with glasses of hot sweet tea. They told me how sorry they were to learn that one of my cities in Australia had just blown away. 'Nonsense' I replied, 'Our Australian cities don't blow away!' 'Yes' they retorted, 'We heard it on the transistor radio, so it must be true!' They were right. During December 24 and 25 Hurricane Tracy destroyed virtually all of Darwin, a coastal city in the tropical Northern Territory of Australia and one I knew very well. But I am getting ahead of my story and must now return to the White Nile.

Eleven kilometres east of the village of Esh Shawal on the White Nile there are two low and symmetrical granite hills called Jebel Tomat or the mountain of the twins. For some weeks I camped between the two hills, and in a pit dug on the col between them found an abundance of Neolithic and younger stone tools, pottery, mammal and fish bones, with large land snail shells in the surface layer. We also found honey from wild bees nesting in a crevice in the rocks, some of which I took as a gift to Omer Mustapha, Umda of Shawal, when he invited me to evening Ramadan breakfast. (I was observing the Ramadan fast because I thought it unfair to eat and drink during daylight hours while my men were fasting and working; it proved a good discipline). He teased the other guests, some of whom were pilgrims on the way to Mecca for the Haj: 'This man is new to our land, yet he brings me wild honey from the mountain!'

A little later I moved camp and accepted the generous offer of Sayed El Hadi el Mahdi to use his private rest house on Aba Island, located 260 km upstream from Khartoum, as my base. A little historical digression is now necessary. Sayed El Hadi Abdel Rahman el Mahdi was the grandson of Muhammad Ahmad the Mahdi who, on the 26th January 1885, finally and decisively defeated the Anglo-Egyptian forces in Sudan led by General Charles Gordon in Khartoum but died a few months later, most likely from typhus. After the Mahdi's sudden death in June 1885, his senior Emir the Khalifa 'Abd Allāhi took charge and ruled the Sudan for the next thirteen years, until the battle of Omdurman on 2nd September 1898, when a combined British and Egyptian army under the command of Major-General Herbert Kitchener, Sirdar of the Egyptian Army, defeated the courageous but outgunned Mahdist army and took control of a country in the throes of famine, disease and social unrest.

By a happy coincidence, the grandson of the Khalifa 'Abd Allāhi was also living on Aba Island, and was a close friend and adviser to Sayid El Hadi, much as his grandfather had been to the Mahdi. One afternoon he invited me for a stroll through the orchards on the island, shaded by groves of mangos, papayas, guavas, grapefruit, limes, oranges, bananas and grapes. As we walked he recited from memory an evocative passage from *The Mahdīya*, the wonderful account by A.B. Theobald of this turbulent era in the history of the Sudan, in which he describes the initial meeting of these two remarkable men. 'As soon as the two met at Mesellamiya, their affection and admiration seem to have been spontaneous. Indeed, each possessed the qualities the other lacked. Muhammad Ahmad [the Mahdi] was a prophet and a preacher; a saint and a visionary; a man who could inspire and charm; one whose own fervour and conviction could make men follow wherever he led. 'Abd Allāhi Muhammad was a man of action and affairs; a fighter and administrator; one who combined a ruthless strength of will with a quick intelligence, and who had the force to translate theories into action. Where the one could inspire devotion, the other compelled obedience.' It was all very inspiring and even somewhat hypnotic. I was beginning to see why so many had given their lives to follow these men over seventy years ago and recall the intense pride with which village elders living along the White Nile dug out battered old tin trunks to show me the chain mail their grandfathers had worn at the battle of Omdurman.

Only a few years later I was to learn of the tragic aftermath of an attempt to re-create history, when a poorly armed militia led by their leaders from Aba Island sought to overthrow the military regime of Major-General Jaafar Nimeiri. Rifles, swords and courage were no match against tanks and bombers. I was in Algeria on a flight from Tamanrasset to Algiers in March 1970 when I read of the death of Sayid El Hadi El Mahdi, shot while trying to escape into Ethiopia. Many of my old friends were to suffer after this, including my friend and mentor the Umda of Shawal who spent some years in prison and was stripped of all except his house, where after his release he played lonely games of cards with those brave enough to be seen with him, until his premature death soon after I last spoke to him in February 1973.

Although I had been very happy leading the life of a reconnaissance soil surveyor among the stoical, good-humoured and hospitable people of the central Sudan, I was becoming restless and felt it was time to move on, specifically to Australia. There were several reasons behind this decision. I wanted to learn more about soils, and Australia led the world at that time in field-based soil science. In addition, I had always enjoyed the laconic humour and relaxed professional style of the Australian soil scientists I had met in Khartoum: Dr Don Drover, Professor of Agricultural Chemistry at Shambat and later Professor of Chemistry at the University of Papua New Guinea in Port Moresby; Tom Smith, an authority on salt affected soils; Dr Robert Smith, our HTS consultant, whom my Dutch reconnaissance soil surveyor friend Jan de Vos used to call 'Dr Boomer', having discovered to his delight that big male kangaroos go by that name in Australia. Perhaps more fundamentally, I was also starting to realise that the physical and chemical properties of the soils I had been mapping for several years in the Blue and White Nile valleys had little to do with present-day climate and even less to do with the underlying bedrock geology. In order to understand how these soils originated and why some were highly saline at shallow depth while others were not, I needed first to understand the depositional history of these two great rivers, which meant knowing the recent geological and climatic history of their headwaters. Ethiopia and Uganda beckoned. As long as I remained fully occupied with the busy daily routine of soil surveys in the Sudan, I would never have the leisure or means to follow the paths opening up to my curiosity. So, in October 1964, I boarded the Greek passenger ship *SS Ellinis* in Southampton and sailed to Australia, intending to stay there for three years.

Chapter 6
Northern Territory, Canberra and Sydney, Australia (1964–1984)

How, then, did this peripatetic Celt, brought up among the lakes and fells of western Europe, but by now familiar also with the Sahara and the Nile, respond to the vast and varied landscapes of Australia? Here I must confess to an early advantage. I had arrived in Canberra in October 1964 to study for a PhD at the Australian National University (ANU) supervised by geomorphologist Joe Jennings. (A geomorphologist is one who studies landscape evolution and the processes that shape landforms. In the UK geomorphology is taught as a branch of geography; in the US as a branch of geology.) Joe and I had already shared a few pints of beer at a country pub in Hertfordshire when I was completing my report on the White Nile soils for HTS. We hit it off immediately. Joe was a bluff and warm-hearted Yorkshireman with a passion for caving, tripe and onions, and Cooper's ale—a South Australian speciality. He had been in the artillery during the last war and had the booming voice to match. Joe was

M. Williams, *Nile Waters, Saharan Sands*, Springer Biographies,
DOI 10.1007/978-3-319-25445-6_6

no narrow specialist, and was interested in every aspect of the Australian landscape, including the glacial landforms of Tasmania, the deserts dunes of central Australia and the coasts of King Island in the south and the Kimberleys in the north. Joe advised me that my arrival was very timely. The CSIRO Division of Land Research and Regional Survey were in desperate need of a geomorphologist and were about to begin a survey in the far north of tropical Northern Territory. Was I interested? Indeed I was. After a short interview with Alan Stewart, Chief of the Division, and with Dr Robert Story, a South African plant ecologist, who was to lead the survey team, they decided that I would do and could I begin next week. (CSIRO stands for the Commonwealth Scientific and Industrial Research Organisation; it is one of the crown jewels of Australian scientific endeavour and is still a great organisation despite savage cuts by a government suffering from invincible ignorance about the nature and merits of good science.)

After some months of careful analysis of the air photos covering the area we had to survey between western Arnhem Land and Darwin (Fig. 6.1), Bob Story and I were ready for action. Tony Hooper, a tough and ebullient New Zealand soil scientist joined us in Canberra. It was by now the dry season in NT so we moved north and began our helicopter survey. A brief bit of background may now be useful. Soon after WW2 the Australian government realised that there was a great deal of land in Australia (and Papua New Guinea) about which they knew very little. Chris Christian and Alan Stewart in CSIRO came up with the concept of 'land system' mapping. A land system was defined as a recurring pattern of topography,

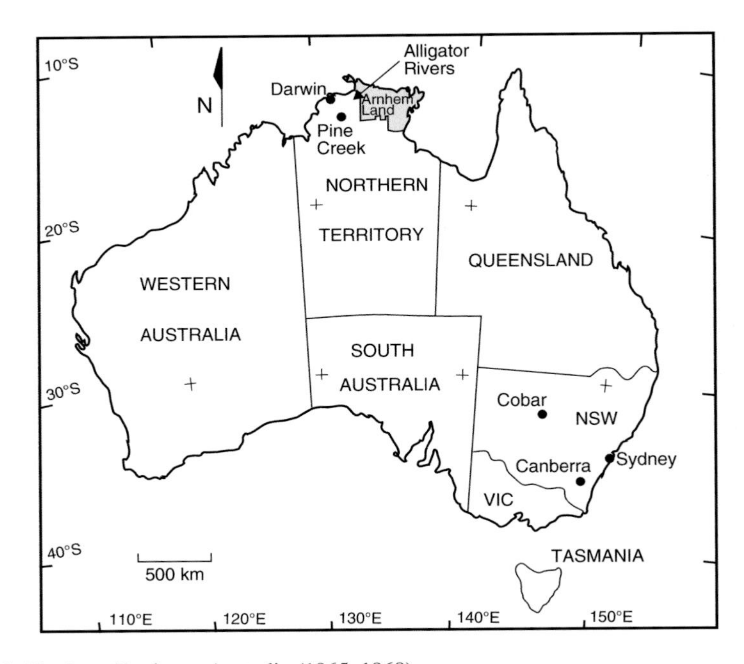

Fig. 6.1 Northern Territory, Australia (1965–1968)

soils and plant cover, and was made up of a series of what were called land units, which were really mappable landforms such as stream channels, hill tops and hill slopes. I have always found the concept of a land system pretty woolly, since it really depended upon the scale at which we were mapping, but it seemed a useful exploratory means of mapping the soils and vegetation covering big chunks of country rapidly, fairly accurately and reasonably cheaply. And so the intellectual effervescence of the Australian National University and the unaffected welcome from my new colleagues in the CSIRO Division of Land Research and Regional Survey soon saw me inducted into the arcane arts of mapping the land systems in forty thousand square kilometres of sparsely populated land between Darwin, Pine Creek and Arnhem Land, including what was to become Kakadu National Park. Many of the research scientists in my old Division of Land Research and Regional Survey moved on to distinguished academic careers as Professors in leading Australian universities, including Mike Norman (agronomy, Sydney University), Wally Stern (agriculture, University of Western Austrlia), Jack Mabbutt (geography, University of New South Wales), Calvin Rose (soil physics, Griffith University), Henry Nix (environmental science, ANU) and Ralph Slatyer (biology, ANU). Ralph later became Chief Scientist to the Federal Government and worked tirelessly to explain and help implement the concept of ecologically sustainable development. We remained in touch for many years.

Perhaps not surprisingly, I became enamoured of the region we called the Alligator Rivers area (Fig. 6.2), and decided that this was the area I would study for my doctoral research. (As an aside, the reptilian denizens of the West, South and

Fig. 6.2 South Alligator River, Northern Territory, Australia

Fig. 6.3 Saltwater crocodile, Northern Territory, Australia

East Alligator Rivers are not alligators but saltwater crocodiles (Fig. 6.3); attaining up to six or more metres in length, they can be lethal for the unwary.) Ever since reading John Steinbeck's stark evocation of the Oklahoma Dust Bowl in *The Grapes of Wrath* I had been sensitized to the perils of accelerated soil erosion. For my thesis I decided to monitor hill slope erosion during three successive wet seasons on low sandstone and granite hills in the area we had previously surveyed. Later, at Joe's suggestion, I enlarged the study to include erosion on sandstone and granite hills in the upper Shoalhaven valley of New South Wales.

I learned a great deal during those years in what the locals call the 'Top End' of Australia, including living off the land during the wet season when a tropical cyclone might turn a shallow creek into a raging torrent three or more metres deep. The local aborigines who wandered past my sporadic campsites taught me how to attach wild cotton to a native bee and run after the bee through the tall *Eucalyptus mineata* and *E. tetrodonta* forest until the bee led us to its nest in a hollow high in a tree. One of the lads then shinned up the tree with a handful of green grass and an aluminium billy-can, scooped out a modest helping of honey ("sugar-bag"), dropped the billy to his companions at the foot of the tree and we then enjoyed a feast of wild honey.

Two Irish tin miners from Roscommon, John and Paddy Toohey, who had been rippers in the Yorkshire coalmines, were watching me dig a soil pit one day, using pick and shovel, which I always enjoyed. They suggested that Alfred Nobel would speed up the process and taught me to use modest amounts of gelignite inserted into shallow boreholes to loosen the soil, which was then easily removed with a long-handled shovel to fashion a trench 2 m long, 1 m wide and 2 m deep. I later completed a course in electrical detonating at the Royal Military College, Duntroon, in Canberra, which I found very instructive. These were the halcyon days before the current epidemic of OH&S rules, which, although good in theory, have, in my humble opinion, become progressively more stultifying and counter-productive as the years go by.

Early one morning towards the end of the dry season I was camped near a small waterhole at the foot of a low granite hill in the Northern Territory of Australia. I was mapping the soils as a prelude to monitoring the movement of soil from the hill slopes once the first rains of the summer monsoon season returned to this region. The soils exposed in my soil pits had aroused my curiosity. On the more poorly drained sectors of the large granite intrusion, the soils consisted quite simply of coarse sands lying directly over the weathered bedrock, but where the seasonal stream channels were entrenched and the slopes well-drained, the soils consisted of a surface layer of coarse sand beneath which there was invariably a layer of quartz gravel up to 50 cm thick, which was in turn underlain by highly weathered granite. As I pondered the possible origins of the stone layer I suddenly noticed how abundant were the grey spiky mounds of the termite *Tumulitermes hastilis* in this locality. I then realised that such mounds had been missing from the swampy portions of the intrusion. I quickly became convinced that the mining activities of the termites during the wet season provided an adequate explanation for the stone layers. The worker termites were able to carry to the surface particles of quartz up to

Fig. 6.4 *Nasutitermes triodiae* termite mounds, Northern Territory, Australia

2 mm in size, which they then cemented in position with a mixture of wet silty clay. Sustained removal of anything finer than coarse sand from the top of the buried weathering front would ultimately concentrate the insoluble quartz gravels within the weathered granite into a stone layer, so that the three-layered soils characteristic of this locality were a result of the mound-building activities of the four main species of termite at this site (Fig. 6.4), and the subsequent abandonment and erosion of the mounds, with rapid loss of clay and silt, leaving behind a surface residue of coarse sand. This meant that conventional models of soil formation simply did not apply here, so that even the individual soil layers could not be described using the ABC system devised in Russia in the late nineteenth century and accepted throughout Europe and North America. A short sharp paper to the *Australian Journal of Science* followed and has been cited ever since.

Two things have always impressed me greatly about the Adelaide Rivers area. One is the widespread evidence of a very long human occupation of this region, extending back in time well over fifty thousand years. The superb galleries of paintings (Fig. 6.5) that adorn the rock shelters and rock faces at the foot of the great Arnhem Land escarpment are but the more recent signs of a human presence, and even these may in some cases date back many thousands of years. The other is the enormous antiquity of parts of the landscape. As the ancient horizontal sandstones of the Arnhem Land plateau erode back, at a rate of about 1 mm a year, they expose even older rocks beneath. The contact with these older rocks is quite irregular and what is being slowly exposed today is a landscape of hills and valleys fashioned over 1700 million years ago. I find this thought quite staggering.

Fig. 6.5 Prehistoric rock art, western Arnhem Land, Northern Territory, Australia

I discovered on one of my return trips to Canberra that Joe Jennings had somehow persuaded the Canberra Times that I was the man to review the 1967 volume on *Landform Studies from Australia and New Guinea*, which he and Jack Mabbutt had edited for ANU Press. I recall his admonition: 'Don't write bloody nonsense!' As an author I can now understand how he felt. The essays in this marvellous book not only alerted me to the enormous variety of research into landform patterns and processes then under way in Australia and New Guinea; they also acted as the catalyst for *Landform Evolution in Australasia* (ANU Press, 1978) edited by Jack Davies and myself as a *Festschrift* for Joe on his notional retirement.

Although I was busily and happily occupied preparing our Adelaide Rivers maps and report, installing my erosion monitoring equipment, and learning from Bob Story how to use a neutron flow soil moisture meter to measure changes in soil moisture at different depths below the surface, my thoughts kept straying back to the Nile. Joe very wisely proposed that I write up my work on the Nile 'to get some practice in writing!' He and Bob Story then surgically dissected my clumsy prose, a humbling but necessary experience. Bob had been senior examiner in English and Afrikaans for the South African Public Service; his scientific papers were a delight to read. He had spent some time with the San people of the Kalahari Desert, spoke their click language as well as Xhosa and excellent German. He later taught himself Spanish in preparation for a year in Patagonia. A master piper, renovator of Baby Austin vintage cars, gold medallist for ballroom dancing, and brewer of first-rate beer, he and his wife Sybil became my surrogate parents in Australia. I could not have wished for a better friend and mentor than Bob. Years later, at Bob's funeral, I quoted George Silberbauer, a South African anthropologist who had known Bob when they both still lived in South Africa. George was no great lover of *Rooinekke* (English-speaking South Africans) but when I mentioned Bob his face lit up: 'Ach! Robert Story—there is a man one is the better for having known!'

While completing writing up my PhD research I had applied for sundry jobs. One day Joe called me in to his office: Jim Rose, a talented New Zealand geographer and head of the School of Earth Sciences at Macquarie University in Sydney, was telephoning to offer me a job teaching soils and hydrology. I asked Jim if I could take leave to spend a few months doing research in the Sahara in the first half of 1970, to which he agreed without hesitation. I liked Jim's relaxed and unpretentious style and accepted immediately, suggesting that I could also teach some geomorphology. We eventually had nearly sixty teaching staff in Earth Sciences with a good spread between geology, geography and geophysics. Field and laboratory work were highly valued, we had an excellent rapport with our students, and a number of the staff went on to develop distinguished national and international reputations for their research. I spent sixteen productive years there between 1969 and 1984, liked and respected my colleagues, and relished the freedom to conduct fieldwork in Africa as well as Australia. Of one such friend and colleague, Dr Gil Jones, the students commented that 'the rocks sang to him'.

It was early in 1969 that I met Don Adamson, a brilliant plant physiologist in the School of Biological Sciences and with his plant physiologist wife Heather a keen caver. Don had an insatiable interest in the natural world and loved crossing discipline boundaries. He soon became a very close friend and valued colleague. He erupted into my office on the fifth floor in Earth Sciences, told me that there was a bright young geomorphology Honours student at Sydney University, one Bob Wasson, working out near Cobar in semi-arid western New South Wales, and sadly in need of proper supervision. My job, said Don, was to get out there and help! I agreed. Bob remains an esteemed friend and colleague.

Don also intimated that he was planning to spend a year teaching at the new Open University at Milton Keynes in England, and from there would be willing to join me for further work along the Nile, if that was what I planned. He was willing to pay his way and we were both perfectly content to live frugally when in the field. Thus began a research partnership that was to last until Don's untimely death in 2002. Don never did manage to persuade me to work with him in Antarctica and on Macquarie Island, where he spent many of his summers. He was one of the very few people in Australia who shared and understood my fascination for the Nile.

Towards the end of 1969 I could once again hear the siren call of the Sahara and so took temporary leave of Macquarie and flew out from Sydney to meet up once more with David Hall in England to launch what we had originally planned as our third expedition to southern Libya. I flew to London via Entebbe, stopping off in Uganda to visit my younger brother Aidan who was teaching at Trinity College Nabbingo located 12 km southwest of Kampala, wandered through the banana groves and over the deeply weathered Buganda Surface, which I found very similar to parts of the Adelaide Rivers area we had surveyed, and explored the excellent library at Makerere University in Kampala.

Chapter 7
Adrar Bous, Central Sahara (1970)

We had been planning a return visit to the southern Libyan Desert for many months and I had been corresponding at length with Professor Desmond Clark about his proposed plans for archaeological work there. Desmond was based at the Berkeley campus of the University of California, and had an unrivalled knowledge of the prehistoric archaeology of Africa. Meanwhile, David Hall had already arranged with a reliable contractor named Katzorakis to transport drums of petrol from Benghazi to Kufra to establish a supply for us there. Then, quite suddenly, the old regime of King Idris was overthrown and a new military regime led by Colonel Muammar Gaddafi came to power. We now had the funds, the personnel, the vehicles and the leave, but our planned destination was no longer open to us. Consternation! By great good fortune, Desmond had been chatting to the French Saharan archaeologist Raymond Mauny, who extolled the archaeological delights

M. Williams, *Nile Waters, Saharan Sands*, Springer Biographies,
DOI 10.1007/978-3-319-25445-6_7

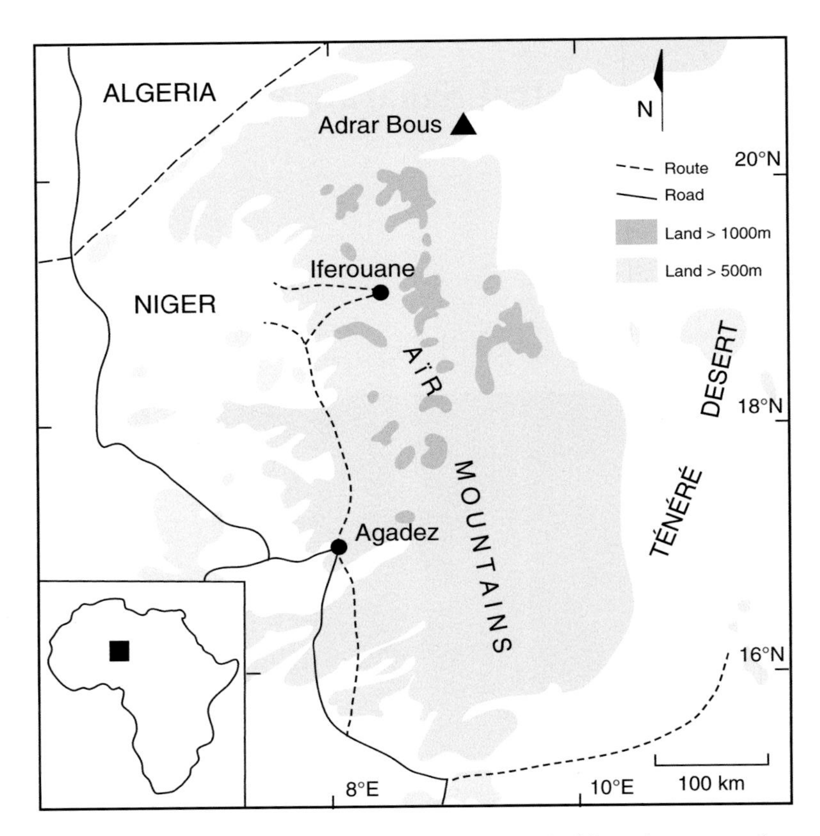

Fig. 7.1 Adrar Bous, Niger, south-central Sahara (1970). *Dashed lines* show route taken across the Sahara

of an isolated mountain in the heart of the Sahara called Adrar Bous (Fig. 7.1). Desmond suggested to David that we go to Adrar Bous instead of Libya, to which we all agreed, with some relief.

Adrar Bous is located 1500 km from the nearest coastline. It lies 65 km east of the Aïr Mts in Niger amidst the dunes and rolling sand plains of the Ténéré Desert. A French expedition using Berliet trucks had paid a brief visit to the mountain ten years earlier, reported fairly abundant surface archaeological finds and had collected some material, but concluded that there was 'no stratigraphy'. By this they meant that all the archaeological material on the surface had been let down from now vanished sediment layers that had been eroded by wind and water, and so nothing was in its original undisturbed state, which made it of little value scientifically. Fortunately, they were wrong. As we found out, some months later, there was plenty of stratigraphy, provided you were willing to dig for it, or to explore the hidden valleys deep within the mountain, where sporadic erosion had exposed good sections in the banks of the ephemeral stream channels.

Fig. 7.2 Valletta harbour, Christmas 1969

Fig. 7.3 Mount Etna under snow, December 1969

I was a member of the small advance party consisting of Tony Pigott, Mike Saunders, John Rogers and John Trewby. They were young lieutenants, tough, cheerful and very on the ball. (They all went on to enjoy very distinguished army and navy careers. Mike and John T. were in the Royal Navy, Tony and John R. in the Royal Engineers. On retirement, Mike was Lieutenant-Commander RN and John T. Rear-Admiral, while Tony was Lieutenant-General Sir Anthony Pigott.) We flew into Valletta, Malta (Fig. 7.2), where we obtained a Bedford RL 3-ton lorry from military disposal stocks. The lorry and four tons of equipment went as deck cargo on a ship from Malta to Tunis. (We later christened the 3-tonner

Fig. 7.4 Tony Pigott near the summit of Mount Etna, December 1969

Josephine for her fits of temperament but she survived two Saharan crossings fully loaded.) We then took the ferry to Sicily with the Land Rovers and buggies we had bought or been given in England, climbed Mt Etna (Figs. 7.3 and 7.4) in a blizzard, drove to Palermo, whence we sailed to Tunis and drove to Carthage to meet David Hall and the rest of our party. Here we endured an unusually cold, wet and muddy New Year's Eve.

It was in Carthage on the last day of 1969 that I finally met Professor John Desmond Clark, little realising at the time that we would work together on many future joint archaeological-geological research missions in Sudan, Ethiopia and India. Desmond was a man of enormous erudition, energy, determination, generosity of spirit and great personal charm, with an unrivalled experience of many parts of Africa. During the war he had served with the British Army's Ambulance Corps in Ethiopia where he was present at the siege of Gondar, which involved walking across a minefield under machine gun fire, and was later part of the Military Administration of Somalia. Throughout his time in Ethiopia and Somalia he proved an indefatigable field archaeologist and collector of artefacts, turning this experience and material into a Cambridge PhD thesis and his first book called the

Fig. 7.5 'The man with the key has gone home.' Passenger truck at Tamanrasset, southern Algeria, January 1970

Fig. 7.6 **a**, **b** Hoggar Mountains, Algeria

Fig. 7.7 **a**, **b** Volcanic plug, Hoggar Mountains, Algeria

Prehistory of the Horn of Africa. It would be easy to write an entire book about Desmond but I shall desist. Suffice to say, he will appear often in this story.

Having collected the 3-tonner and our equipment from the docks at Bizerta near Tunis we drove through the green hills of northern Tunisia and Algeria before

Fig. 7.8 **a**, **b** Andrew Warren at Assekrem, Hoggar Mountains, Algeria, January 1970

crossing the Atlas Mountains to enter the Sahara proper. The exceptionally heavy rains we had endured at Carthage were to dog us all the way south past the oasis of Biskra and as far as the date factory at Touggourt, to our minor discomfort but to the delight of the local people. From then on we moved into an arid landscape of vast gravel plains, great volcanic mountains, scattered dunes and isolated sandstone plateaux, reminding me of earlier days in Libya. We drove down through Arak Gorge and on to Tamanrasset (Fig. 7.5), a small town near the foot of the mighty Hoggar Mountains (Figs. 7.6a, b and 7.7a, b). While we were immobilised in Tamanrasset awaiting customs clearance (they thought our sacks of dried spinach were some illicit drug and were deeply suspicious of our intentions), Andrew Warren (a friend from my university days, who had studied the dunes of western Sudan) and I drove up to the hermitage of Charles de Foucauld on the summit of a flat-topped volcanic mountain called Assekrem (Fig. 7.8a, b). Charles de Foucauld—a former officer in the French army in Algeria—became a hermit in 1905, was revered by the local Tuareg people but was murdered by Tuareg raiders from outside the region in 1916. Inside the stone hermitage at Assekrem we found a copy of Pierre Rognon's

Fig. 7.9 Faulted rock face, Aïr Mountains

recently published doctoral dissertation on the geology and landforms of the Atakor, a remarkable piece of work, which I later read in detail, before unexpectedly meeting Pierre during a conference in December 1971 held in Addis Ababa, Ethiopia.

Released from Tamanrasset, we drove on into Niger and down to Agadès at the southern tip of the Aïr Mountains. Here, in the Hotel de l'Aïr, we met some French colleagues, including Henri Lhote, who had written so vividly about the Tassili frescoes, with their vast galleries of prehistoric rock art. We set up a supply depot in Agadès and our party now split into two main teams. The dune party with dune specialist Dr Andrew Warren set off into the Ténéré Desert to study dune forms and processes; and the Adrar Bous team set forth north through the Aïr Mountains (Fig. 7.9) in three Land Rovers as far as Iferouane, a small oasis in the heart of the mountains, after a drive along very rough tracks. Our Land Rovers suffered; two of the half-shafts broke a few hours out from Iferouane, depriving us of four-wheeled drive. David wisely decided that all the vehicles had be nursed back to Agadès for repairs, so Desmond and I set forth on January 21 with three camels guided by Zewi bin Weni, a local Tuareg and former *goumier* or desert military scout, who knew the region well. We travelled light: a few jerry cans of water, some sticks of wood, some army 'compo' rations, and our personal gear. It took us three days to reach the mountain, where Zewi left us and turned back to collect Desmond's two archaeology graduate students, Andy Smith and Alan Pastron, who arrived some days later, with Ibrahim, a pleasant lad from Agadès whom we had hired as a cook. Andy hailed originally from Glasgow; he had served in the U.S. Air Force, was widely travelled, very practical and a brilliant excavator. He later worked among the Tuareg on Neolithic sites in the Tilemsi valley of Mali before moving to a teaching post at the University of Cape Town. Alan was less worldly wise than Andy, and sometimes found conditions trying. Most evenings at Adrar Bous he and I played chess on a full-sized board I had brought along. We were evenly matched, which made for some exciting contests. Also with us for a while was Dr Peter Beighton, our medical doctor, as ebullient as ever. He and a local Aïr Tuareg later walked across the desert to Tamanrasset, in an interesting study aimed at comparing water intake and output between a local man of similar physique and age to Peter and

Fig. 7.10 a, b Sand dunes east of the Aïr Mountains, Niger, south-central Sahara, January 1970. *Photo* J.D. Clark

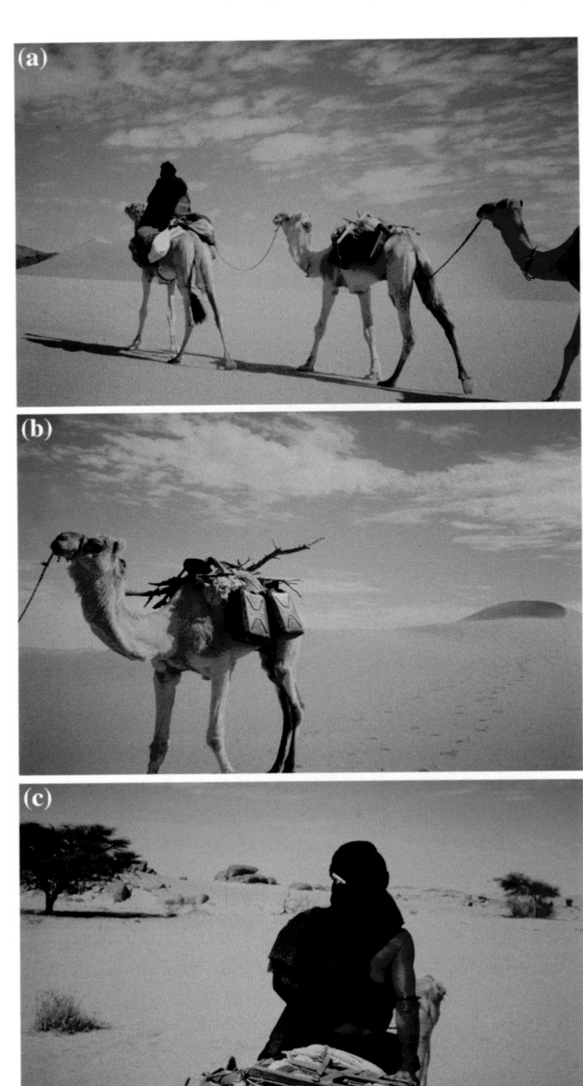

Fig. 7.11 a–c Zewi Bin Weni, our camel guide on the way to Adrar Bous, January 1970

accustomed from birth to the desert and a fit but only recently acclimatised European. The result: no difference!

Soon after Desmond, Zewi and I had moved out from the mountains and valleys of the Aïr massif astride our three camels (Figs. 7.10, 7.11 and 7.12), we found ourselves in a very wide flat valley, bounded to the west by the massif and on the east by a great sand dune nearly a hundred metres high, from the lower flank of which emerged a horizontal river terrace made up of finely layered alluvial silts and

Fig. 7.12 Desmond Clark riding his camel on the way to Adrar Bous, January 1970

Fig. 7.13 Neolithic and Mesolithic stone tools seen on the camel trip to Adrar Bous

clays. A river terrace is an abandoned flood plain and so this terrace was clear evidence that water had once flowed across this now dry valley floor. On the surface of the terrace we found stone tools with a likely age of about 12–15,000 years. Upon leaving this enclosed valley floor we moved out onto the undulating sand plains leading to Adrar Bous. As we approached our destination there were more

Fig. 7.14 Neolithic pottery west of Adrar Bous, January 1970

and more stone tools littering the surface (Fig. 7.13) together with fragments of pottery, so that Desmond's finely honed archaeological senses became even more alert. At one spot we found exposed at the surface the sand-blasted base of a pot that had been inverted and used as a cover for another more ornate pot (Fig. 7.14) filled with the desiccated fruits of a tree (*Celtis integrifolia*) requiring about 500 mm of mean annual rainfall. Beneath the pot was a Neolithic human burial, which we left in place.

Once we had reached Adrar Bous we began a systematic exploration of the mountain and began to excavate promising sites. We ultimately found stone tools ranging in age from about 500,000 years ago onwards. My studies of the sediments and ancient soils with their associated stone artefact assemblages allowed me to reconstruct a succession of wetter and drier climatic phases culminating in the still well preserved evidence of former lakes near the southern margin of the mountain. Some words of explanation are necessary here.

Adrar Bous is what geologists call a 'ring complex'. From the air a ring complex looks a bit like a bull's eye, and can range in shape from circular to elliptical. The different rings of rock that make up the ring complex can vary a great deal in their resistance to weathering and erosion. As a result the more resistant rocks stand out as roughly concentric ridges while the less resistant rocks form concentric or crescent-shaped valleys. During times in the geologically recent past when the climate was wetter and the regional water tables were at quite shallow depth, small lakes sometimes occupied the valley floors. These lakes were fed by a combination of runoff from the nearby uplands, direct rainfall onto the lakes, and subsurface flow from the groundwater. Nile perch, hippos, crocodiles, turtles and aquatic snails lived in these lakes and I later obtained radiocarbon ages for the shells of these snails.

And so the weeks went by, with occasional sand storms to try us and fill in an old well I was trying to open up, the chance discovery of another beautiful saw-scaled viper early one morning, and one evening the unforgettable sound of what early French Saharan explorers called *le tambour des dunes* (the drumbeat of the dunes). This extraordinary noise is caused by a layer of sand avalanching down the slip-face of a high dune. It probably terrified the early inhabitants of these regions, and was certainly enough to halt conversation and make us listen.

On a cold and windy morning shortly after dawn, Desmond and I were walking pensively along the shore of a lake that we later established had dried out some nine thousand years earlier. It was our last day at Adrar Bous. Desmond had spent much of his professional life immersed in the prehistory of Africa, but this had been his first visit to the Sahara. The day was March the 20th 1970 and was the final morning of nine weeks spent investigating the prehistoric archaeology and recent geological history of this remote and isolated spot. We suddenly noticed some bones that the wind had exposed from their former protective cover of ancient lake silts. The bones belonged to a hippopotamus skeleton. Embedded in the rib cage was a barbed bone harpoon point of the sort widely made throughout the southern Sahara and Nile valley roughly eight thousand years ago. The chance discovery, aided by the deflating action of the constant desert winds, took us both back in memory to a similar early morning stroll nine weeks earlier, the day following our arrival at Adrar Bous after a three day camel trip from the Aïr Mountains, when we

Fig. 7.15 Five thousand years old short-horned Neolithic cow skeleton (*Bos brachyceros*) excavated at Adrar Bous, south-central Sahara, February 1970

Fig. 7.16 Proto-historic rock engraving, Aïr Mountains, south-central Sahara

had spotted the white trace of a horn core embedded in dark brown clay close to our campsite. The horn core proved to be that of a short-horned Neolithic domestic cow, *Bos brachyceros*, which had died in a small pond over five thousand years earlier (Fig. 7.15), and whose complete skeleton had remained intact once the clays dried out and set like concrete. It eventually took fourteen days with dental picks to excavate the remains.

Having spent nine weeks investigating the geology and prehistoric archaeology of Adrar Bous, Andy Smith and I set forth by camel from the oasis of Iferouane to explore the rock art of the northern Aïr Mountains. In this enterprise, Mamunta ben Tchoko, a local Tuareg who knew the mountains well, proved an excellent guide. One remarkable engraving showed a man with three plumes on his head leading either a small horse or a baboon (Fig. 7.16). Another, alongside a deep underground pool of water, depicted a gerenuk antelope, a type of antelope no longer present in this region. Other engravings showed elephants, giraffes and other large savanna mammals. The engravings thus provided us with a window into a series of former worlds, with climates quite different to the aridity of the present-day. Three thousand kilometres to the east, at Jebel 'Uweinat, astride the borders of Egypt, Libya and Sudan, there are superb polychrome paintings of great herds of domestic cattle, further testimony to a time, however brief, when the Sahara was less arid and host to numerous small lakes. It is hard to imagine a greater contrast between a lake-studded Sahara and the present waterless wilderness.

One evening, after a very long day, we stopped to unload the camels, prepare our evening meal and settle down for the night when we discovered that our only small sack of rice was missing. We concluded that it had not been properly attached and must have fallen to the ground during the day. The camels were pretty tired, as were we. Mamunta promptly fished out from one of his leather bags several copious handfuls of an interesting green leaf and gave it to his camel. He himself chewed some of the leaves and fashioned them into the size of a golf ball, which he placed in one nostril of the camel. The camel rose up, eyeballs bulging, shook his head, scattered green spume everywhere, and roared that he was ready for action. The bag of rice was duly retrieved; it had fallen off early in the day.

During my time at Adrar Bous I had learned a great deal about African prehistoric archaeology from Desmond, much of it very practical. One day he suggested that we should prepare a joint research proposal to the U.S. National Science Foundation to fund a three-year study of the prehistory of Ethiopia. I agreed and suggested that we could also do some joint work in Sudan: I could provide a Land Rover, look after the logistics and show him the sites I had discovered in the lower White Nile valley. Desmond had written about early agriculture in the Nile valley but had never actually worked there, so he thought that some joint work there was a good idea. He also urged me to present the results of my Adrar Bous work at a conference to be held in Addis Ababa in December 1971. This conference went by the grand title of the Pan-African Conference on Prehistory and Quaternary Studies, and would be an excellent opportunity to meet many of the leading figures in this field working in Africa. The Quaternary Period spans the last 2.6 million years and is the chunk of geological time that has always interested me the most, since it has such well preserved evidence of past fluctuations in environment, including climate. Again, I readily agreed. The return to Australia allowed time for a brief return to Uganda and a few days to climb Mount Kenya.

Chapter 8
Ethiopian Highlands and Rift Valley (1971–1978)

Following Desmond's suggestion, I did attend the Pan-African Prehistory conference, which was held in Addis Ababa, Ethiopia, in December 1971. The Emperor Haile Selassie (who was nearly eighty years old at the time) presided at the opening ceremony in Africa Hall with its magnificent stained glass windows. There was a grand barbecue that evening with spit-roasted oxen, at which I first tasted *tej*, the delicious Ethiopian mead made from honey. The emperor later held a reception for us in his palace. We were led up one at a time to exchange a few words with him in English or French as he sat on a throne in the reception hall fondling a small dog. He had in his earlier days done great things for his country, but in his old age he did not seem to be well informed about events outside the narrow confines of the palace. In particular, he did not seem to be aware of the tragic impact of the severe 1973 famine in Wollo province. Popular unrest swelled and he was eventually

M. Williams, *Nile Waters, Saharan Sands*, Springer Biographies,
DOI 10.1007/978-3-319-25445-6_8

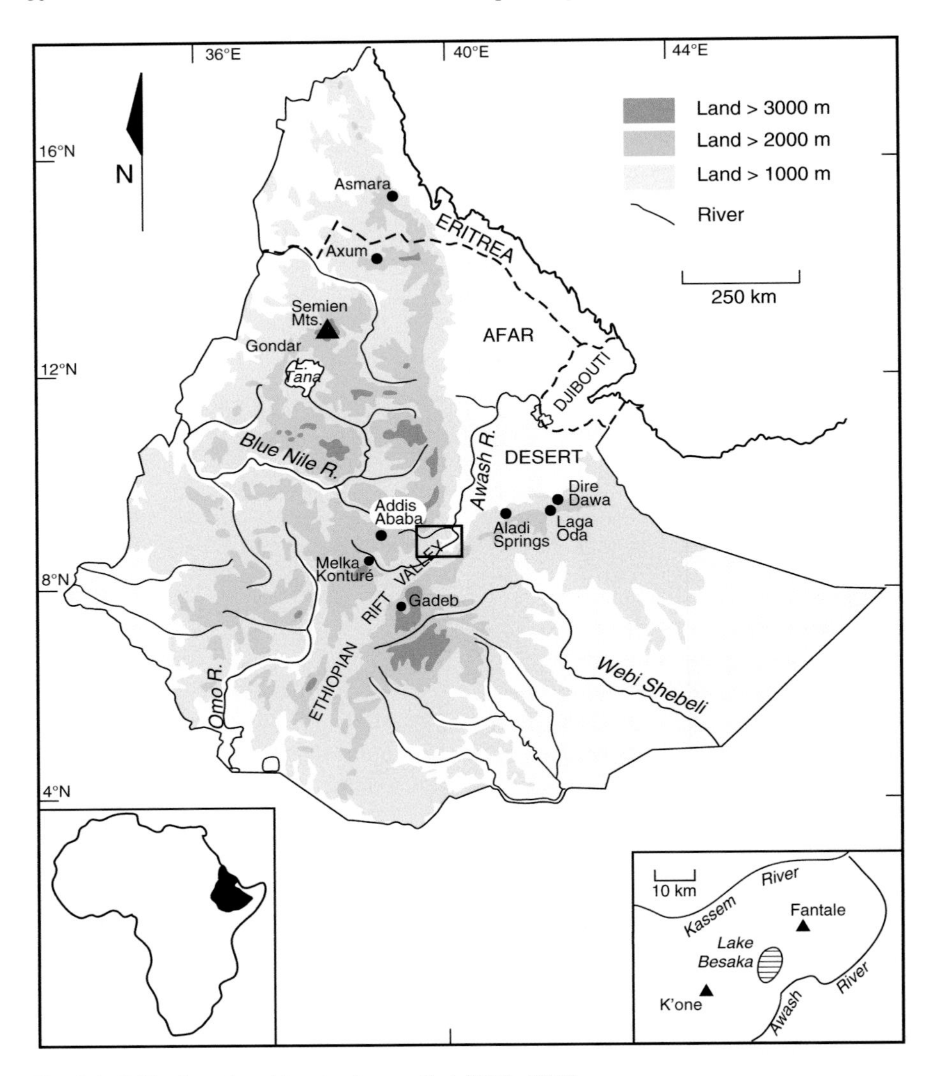

Fig. 8.1 Ethiopia and prehistoric sites studied (1971–1978)

obliged to abdicate in September 1974 and died a year later, a victim of Mengistu Haile Mariam's murderous regime. Mengistu himself was forced to flee to Zimbabwe in 1991, at least in part because of severe unrest triggered by another prolonged famine. But I digress.

Apart from the conference lectures, there were three very interesting field excursions. One was a day trip to the site of Melka Kontouré located 50 km south of Addis Ababa (Fig. 8.1) close to the Awash River, where Jean Chavaillon and his wife were excavating an Early Stone Age site replete with plenty of fossil mammal bones. Here Mary and Louis Leakey threw out a number of useful ideas, as did Desmond Clark and Glynn Isaac.

Glynn was working out at Olorgesaillie in the Kenya Rift Valley and at Lake Turkana in northern Kenya with Richard Leakey, and we met up a year or so later in Sydney when he came over to Australia to give some talks and to visit his twin brother, who was teaching American history at Latrobe University in Melbourne. (Many years later, after Glynn's untimely death and when Betty and Desmond Clark were staying with us at our home in Mornington, a seaside hamlet southeast of Melbourne, we all set forth to visit Glynn's parents in their home among the sand dunes at the end of the Mornington Peninsula. Glynn's father had been born in Barry, went to school with my mother, insisted on addressing me in Welsh, to my intense embarrassment, and by another coincidence had known my former much loved CSIRO boss, Bob Story, in Grahamstown, South Africa, where both had been competing for the same Chair of Botany.)

The elderly Raymond Dart and his wife were also present at the Melka Kontouré excursion. As a young and iconoclastic Australian Professor of Anatomy at the University of the Witswatersrand in Johannesburg, South Africa, Dart had identified a new species of small-brained hominid, which he named *Australopithecus africanus* in a letter to the leading international scientific journal *Nature* published in 1925. His paper caused a furore among the British fossil establishment, who had been misled by the infamous 'Piltdown Man' hoax (finally exploded by the British geologist Kenneth Oakley in 1953) into believing that the earliest ancestral Englishmen had a big brain, as one might have expected! It was also on this short excursion that I first met Hugues Faure, a brilliant and very fit French geologist who had carried out excellent geological investigations in the deserts of Niger and who was now working in the Afar. His nickname was *le guépard* ('the cheetah') from his habit of running from one geological section to another under a burning sun.

Another short excursion was down to the delta of the Omo River in southwest Ethiopia, where rival archaeological teams from three nations (France, Kenya, USA) were working hard to find the earliest ancestral human fossils. Yves Coppens, a highly experienced and urbane French palaeontologist (fossil expert), was leader of one of the teams and explained very clearly the nature of the fossil finds and their geological setting. My most vivid memory is not so much of the fossils, impressive though they were, but of the brown and sluggish waters of the Omo and the sheer numbers of crocodiles along the banks of the river and in the water. The adjective 'pullulating' best describes their abundance. I suspect that during the two rainy seasons when the river is in flood the crocodiles can scatter more widely in search of food, but as the river shrinks during the intervening dry seasons, they are forced into a more confined space.

The third excursion was down to the Afar Desert and was led by the energetic, generous and exuberant French geologist and palaeontologist Maurice Taieb. In this harsh and arid volcanic landscape Maurice had discovered many hundreds of fossils ranging from a few million to a few thousand years in age. Maurice was keen for others to join him, and he later worked jointly with the American palaeontologist Don Johanson, whose team gained fame in 1976 with the discovery of the hominid

Fig. 8.2 Semien Highlands, Ethiopia: dissected 30 million-year basalt flows near headwaters of Blue Nile

Fig. 8.3 Frances Dakin and game guard, Semien Highlands, Ethiopia, 1975

fossil 'Lucy', subsequently proposed by Johanson and Tim White as the type fossil for the new hominid species *Australopithecus afarensis*. Karl Butzer and his wife Elizabeth also came on the Afar excursion. Although he and Maurice did not always see eye to eye, I liked Karl and enjoyed his sardonic sense of humour. He was a brilliantly perceptive field stratigrapher and had an encyclopaedic knowledge of prehistoric archaeology and Quaternary geology.

All in all, the conference and associated field excursions proved a heady combination and gave me plenty to think about. With still some weeks of leave in hand, I wanted to discover more about Ethiopia and was heading off to climb Ras Dashan in the Semien Mountains (Figs. 8.2 and 8.3), when Bill Morton invited me to join him and Dr Getaneh Assefa on a student fieldtrip to the Blue Nile gorge. Bill and Getaneh taught geology at what was then Haile Selassie I University and is now the University of Addis Ababa. I suggested to Bill that if we could date the uppermost lava flows in the gorge we could obtain a maximum age for the gorge and also work out a mean rate of geological erosion. Ian McDougall at the Australian National University was ultimately able to determine credible potassium-argon ages for the basalt flows and we duly published the results in *Nature*.

After climbing Ras Dashan I set forth by bus for Asmara via Axum, the ancient religious capital of Abyssinia. We were driving along what was then a narrow dirt road cut into the mountainside between Gondar and the Blue Nile gorge when a man suddenly appeared from the bushes and walked straight into the path of the oncoming bus. The driver braked hard but unfortunately the bus struck the man. The men in the bus, all of whom carried WW2 Italian rifles, wanted to throw him over the edge of the cliff, before his friends arrived to extract vengeance. Both the driver and I refused. He had a broken femur and some abrasions, but was otherwise fine, so I patched him up and we loaded him on the bus, bound for the nearest hospital. At the first village we disembarked and I found a dressing station. Meanwhile the police ordered the hapless driver to take them to the scene of the accident, no doubt to extract money from him. Having purloined a stretcher from the now empty police station by climbing through a window, we took our patient to have his leg splinted. He recovered consciousness briefly, saw my blue eyes and ginger beard just above his face, and promptly passed out once more. His father then arrived, to thank me for helping to save the young man's life. It transpired that he had been high on some form of local narcotic or hooch and so was quite relaxed when he walked into the oncoming bus, which may have saved him from more serious injury. The disconsolate driver now returned and we set forth anew with our wounded patient. It was now nightfall. At the bridge across the Blue Nile gorge the police proved awkward and refused to allow us passage on the grounds that driving at night was forbidden. I was angered by their attitude. The male bus passengers were by now fully committed to getting the injured man to hospital and surrounded the police who, outnumbered and outgunned, saw sense and let us through.

From Asmara I flew to Kassala in Sudan, caught a bus to Khartoum, and arrived in time to greet Don Adamson as he was alighting from his early morning Swissair flight. We completed some useful fieldwork on both sides of the White Nile (see Chap. 9) and I then flew back to Asmara and caught local buses back to Addis Ababa. On the way I stopped off in Axum and the monks in charge of the crown jewels very kindly brought them out into the sunlight so that I could admire them more closely and take photographs (Fig. 8.4). I doubt that would be possible today. From Addis I caught the train to Dire Dawa, explored the beautiful limestone country around the town, and then travelled in the goods van to Djibouti in what was then the French territory of TFAI (*Territoire Français des Afars et des Issas*).

Fig. 8.4 Ethiopian crown jewels, Axum, January 1972

(The Afars were the Afar or Danakil tribesmen of the Afar Desert and the Issa were the Somali nomads of that region; both were constantly raiding each other for cattle and both had a well-deserved reputation for ferocity.) Since I was by now seriously short of cash, I deposited my backpack at the railway station, stocked up on oranges and water, and spent a couple of nights sleeping on the beach close to the Bishop's palace. I was awaiting the arrival of *SS Patris*, a Greek migrant boat and sister ship to *SS Ellinis*, on which I had first sailed to Australia in 1964. On my final night I was fast asleep in my hole in the sand when a man fell on top of me and began to attack. We struggled for a while until we found a common language: Arabic. He was a Yemeni fisherman in exile from the regime of Prince Badr and was terrified that I was some drunken cutthroat Scandinavian sailor. He then led me to the main road in Djibouti and proposed that the roundabout would be safe. I thanked him, and once he was gone, I returned to my spot on the beach.

December 1971 and January 1972 had given me the chance to see a little of Ethiopia and I had been entranced by the beauty of the land and its people. I was therefore eager to resume my acquaintance with this mysterious land of high mountains, deep valleys and hidden churches. During the first two months of 1974 and of 1975, when the rains had ceased in the Ethiopian Rift Valley, I joined up with Desmond and Betty Clark and the team of Berkeley graduate students in African archaeology, and resumed the type of work I had begun with Desmond at Adrar Bous in 1970. My job was to use the evidence left by sediments and fossil soils to reconstruct past changes in prehistoric environments, including climatic changes It was the type of forensic detective work I enjoyed and involved crossing the artificial boundaries between many different disciplines, which always appealed to me.

Fig. 8.5 Lake Besaka in January 1974 with Fantale volcano in the background, Ethiopian Rift Valley. The lake level has risen several metres since then

Fig. 8.6 Lyre-horned Karriyu cattle, Ethiopian Rift Valley, January 1974

We began work at Lake Besaka in the Ethiopian Rift, a magnificent location overlooked by the great gaunt volcanic mass of Fantale, with its fumaroles and historic lava flows (Figs. 8.5, 8.6 and 8.7). When we started work at Besaka in 1974 the lake waters were brackish and host to great flocks of pink flamingos. Next year my levelling showed a 50 cm rise in lake level and the flamingos had gone. The rising of the lake continued, flooding the main road and railway from Addis to Dire Dawa, which had to be rebuilt on raised embankments. No one knew why the lake level was rising. Was it a resurgence of hot spring activity, often a precursor to a

Fig. 8.7 Dust storm in the Awash National Park, Ethiopia, January 1974

volcanic eruption? Was it illicit diversion of water from the Awash River to irrigate the Metahara sugar cane plantations? We had worked out in some detail the history of Lake Besaka for the last 15,000 years and now saw before our eyes the re-creation in one year of the lake of 11,000 years ago, when the climate was wetter throughout the region and all the Ethiopian Rift Valley lakes were high.

The local Karriyu cattle herders did not understand what we were doing and did not appreciate our presence. They threatened to burn our vehicles and shoot us. It was in fact a colossal bluff, as the wife of one of the cantankerous Karriyu whom I was treating for malaria told me, but the local governor came and explained what we were up to. I once asked two young Karriyu lasses why they did not get their husbands to help carry the heavy goatskin bags of water they ferried each day from a couple of freshwater springs near the lake to their camp. They snorted: 'ask the men to carry these bags? Why, they can't even lift them!'

After excavating several Late Stone Age sites near Lake Besaka we moved camp to K'one (previously 'Garibaldi') volcano, and camped inside one of the many nested craters within that volcano. On a nearby hillside we found a prehistoric obsidian quarry site and could reconstruct the precise sequence with which some of the stone tools had been made. (Obsidian is black volcanic glass and can be wonderfully sharp; the Karriyu wanted me to use obsidian blades to shave my beard. I declined.) What I found intriguing about the K'one site was that the layers of horizontal sediments on the floor of our crater were cut by a dense network of now stable gullies that appeared to have been triggered by an earthquake that took place about 5000 years ago and created a deep fissure in the rocks ('welded tuffs') lining the floor of the crater. This great crack then diverted runoff and sediment underground. The government of the day was blaming local charcoal burners for causing the gullies so I gave a talk in Addis and published a short account in the *Ethiopian Journal of Science* (*Sinet*) to explain that this was not so.

Fig. 8.8 Desmond Clark examining Middle Stone Age fossils sealed beneath a stalagmite, Porc Epic Cave, Dire Dawa, Ethiopia, 1975

Fig. 8.9 View from the Porc Epic cave across the limestone hills near Dire Dawa, 1975

We moved east out of the main Ethiopian Rift Valley and excavated another site at Aladi Springs on the southern margin of the Afar Rift. Desmond and I did a rapid reconnaissance and located a limestone cave site situated about 150 m above the floor of a wadi near Dire Dawa (Figs. 8.8 and 8.9). Getting to the cave involved cutting our way with pangas (machetes) through a dense thicket of prickly pear, which showered us with tiny sharp spines. The spines embedded in us soon turned septic and for days afterwards we were both prising the spines out of our anatomies. A formidable trio of renowned French priest-scientists (Abbé Breuil, Père Teilhard de Chardin and Père Paul Wernet) had excavated this cave nearly half a century earlier. It was known as the *Porc-Épic* (or Porcupine) Cave, doubtless from the abundance of porcupine quills formerly present on the floor of the cave. Middle Stone Age hunters had occupied the cave seasonally and had left behind the bones of the animals they had slain and eaten, together with their spears, only the stone points of which survived.

One afternoon during the time of the troubles in Addis Ababa we had a spot of bother. I had climbed down to the wadi floor with an empty 20-l jerry can (which I used to carry full of water each morning up to the cave) when I saw a local tribesman trying to strangle one of our Amhara workers who came originally from the highlands. It was a time to settle old scores. The tribesman was red-eyed and frothing at the mouth. I stopped to consider the best way to separate the two when Desmond appeared quite suddenly. With a stentorian: 'Now look here my good man, what do you think you're doing?' he prodded the tribesman in his belly with his trusty walking stick. The man looked down in horror at this secret weapon … and promptly fled. Desmond invariably carried a stout walking stick with a metal tip to poke at possible stone artefacts on the ground as he walked along.

On another occasion we had set up camp near to the painted rock shelter of Laga Oda among some rugged limestone hills and I had spent the day exploring for possible cave art. As the sun was setting I was making my way back to camp along a narrow ledge bounded by a steep cliff above and below. Quite suddenly a male and female leopard appeared from around a corner, walking in single file. There must be a hard-working guardian angel looking after idiot earth scientists like me because for some reason I imagined they were cheetahs. Just as well. I admired these two beautiful creatures while they gazed at me, knowing that cheetahs were pretty harmless and forgetting that they like the open plains and don't frequent rocky cliffs, unlike leopards. The leopards then turned and went back the way they had come. I returned to camp and was telling Desmond about my encounter with

Fig. 8.10 Oromo men and women watering their horses in the Webi Shebeli, Gadeb, Ethiopia, 1975

Fig. 8.11 Desmond Clark, Frances Williams and Alamayu looking across the Gadeb plain, 1978

the two beautiful cheetahs up on the cliffs above our camp. He gave me a strange look: 'No, Martin, leopards, not cheetahs!' and I realised then that he was right.

From the lowlands of the southern Afar Desert we moved up to the high grassy plains of Gadeb at 2300–2400 m elevation in the southeast highlands of Ethiopia. Desmond, Frances Dakin (a geologist at the university in Addis) and I had carried out an exploratory survey of Gadeb in 1974 and had found plenty of good sections from which protruded Early Stone Age hand-axes and cleavers. We also found it exceedingly cold and windy and used dried pats of cattle dung as fuel to cook a hot and reviving meal at night.

The wide waters of the Webi Shebeli (Fig. 8.10) meander across the Gadeb plain before vanishing into a deep gorge and eventually flowing on to Somalia, never to reach the sea. The present-day flood plain of the Shebeli River is very level. This is because a lake once occupied the valley and fine sediments were laid down on the floor of the lake, filling in any hollows in the older rocks. About 2.71 million years ago a volcano erupted and a lava flow dammed the river, forming a lake 30 km long, 10 km wide and up to 40 m deep. (In December 1975, accompanied by Pierre Rognon and Françoise Gasse, two French earth scientists with whom I had just been working in the Afar Desert, I had the chance to locate the dam, which we later sampled for potassium-argon dating by Garniss Curtis, a friend and colleague of Desmond's at Berkeley.) The lake persisted for roughly three hundred and fifty thousand years, at which time the lava dam was breached and the lake drained, cutting a deep gorge downstream. After the lake had drained and the river resumed its activity early humans (*Homo erectus*) moved in and used the abundant river gravels to make stone tools and hunt small game. Our discovery of the bones of a hippo in association with stone tools suggests that they may have butchered any recently dead large animals they came across. There is some evidence from our work that they may also have been using fire and bringing in obsidian from at least

Fig. 8.12 Françoise Gasse and Mike Tesfaye near the Webi Shebeli gorge, 1978

Fig. 8.13 Don Adamson inspecting an Early Stone Age site at Gadeb, with the author standing nearby

a hundred kilometres away. Other members of our team included our Ethiopian Antiquities Officer Alemayu (Fig. 8.11), who had earlier discovered some magnificent hominid jawbones in the Afar desert, Raymonde Bonnefille (an authority on Ethiopian fossil pollen) and my good friend and colleague Françoise Gasse (Fig. 8.12), who carried out detailed studies of the Gadeb lake history using fossil diatom remains. Don Adamson (Fig. 8.13) was also an energetic member of the team.

The clouds of unrest were building up and in the capital there were constant skirmishes between what were called the Red Terror and the White Terror brigades. It was during one such turbulent time in 1978 that Bill Morton, a close friend of both Frances and mine, was killed by the militia on the outskirts of Addis Ababa. After the event, Bill was accused of being a spy and of hiding vast sums of money in his socks to aid anti-revolutionary elements. In the many years we had known Bill he wore sandals and never wore socks… Bill's death was a tragic loss of a brilliant geologist and superb teacher who taught and inspired many young Ethiopian geologists.

Chapter 9
Back to the Sudan: The White Nile Valley and Jebel Marra Volcano (1973–1983)

In early January 1972, after leaving Asmara (in what was then Ethiopia and is now the capital of Eritrea), I had flown to Kassala in eastern Sudan, and briefly explored its spectacular granite mountain, which I had last seen a decade earlier. I then caught a bus to Khartoum in time to meet Don Adamson at the airport. We purchased some field equipment, hired a car and driver and set forth southwards along the east bank of the White Nile. On the way to the village of Esh Shawal (Fig. 9.1) we collected sand samples down to depths of 2 m in the sand dunes that extended over a north-south distance of nearly 150 km east of the present river. I wanted to test an idea I had long held that the sands that made up these dunes had originally been brought down from the highlands of Ethiopia by the Blue Nile and later ferried

M. Williams, *Nile Waters, Saharan Sands*, Springer Biographies,
DOI 10.1007/978-3-319-25445-6_9

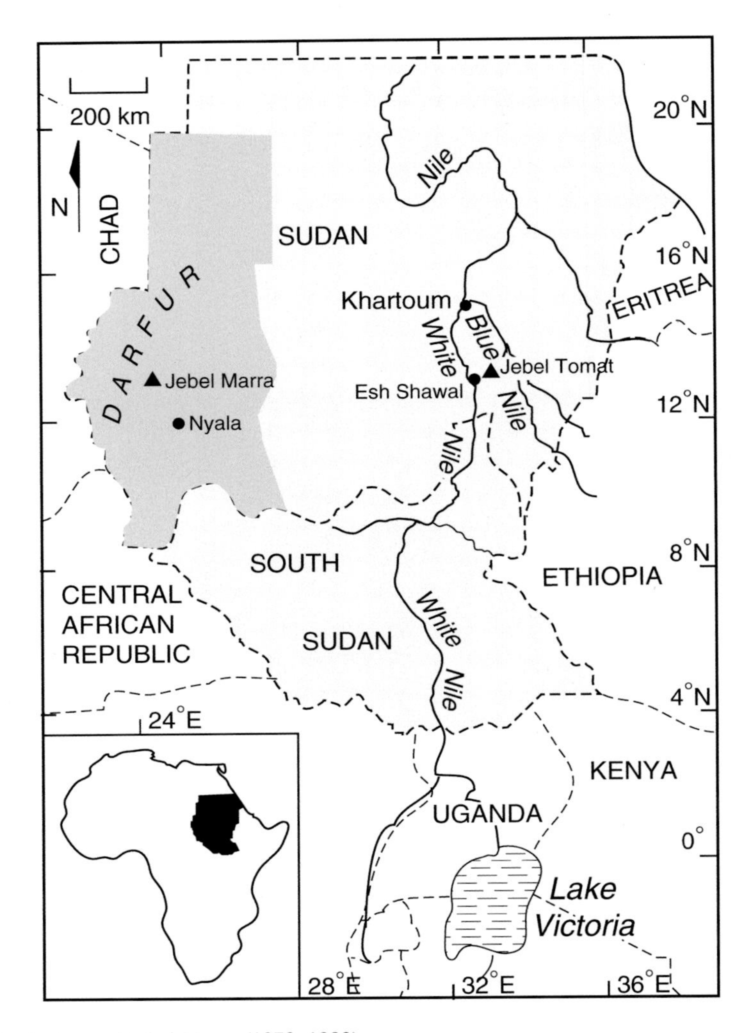

Fig. 9.1 Sudan and Jebel Marra (1972–1983)

by a network of once active channels that radiated across the plains of the northern Gezira until they reached the White Nile. Here the sands were blown out from the channels to form dunes. By analysing the heavy minerals present in these dune sands, we would be able to determine their initial geological source area, which did indeed prove to be the volcanic uplands of Ethiopia.

We also took a short trip up the White Nile with biologist Dr Asim el Moghraby on the University of Khartoum Hydrobiology Research Vessel RV Malakal in order to examine the banks of the White Nile and collect sediment samples from the bed of the White Nile (Fig. 9.2).

Fig. 9.2 Dr Asim el Moghraby and the author collecting sediment samples from the bed of the White Nile, January 1973. *Photo* Don Adamson

A short recapitulation is now needed (see Chap. 5). During 1963–64, while mapping soils east of the lower White Nile in central Sudan, I had described and sampled 556 inspection sites, of which 440 were bores and 126 were pits. One pit, near the village of Tagra, contained freshwater snail shells at a depth of a metre. These shells proved to be 8000 years old by radiocarbon dating; another pit, near Esh Shawal village, had aquatic snail shells that were even older, approaching an age of 15,000 years. Now, a decade later, my field companion Don Adamson and I re-opened and widened the Tagra pit, finding two shell beds, between which Don later recovered a barbed bone harpoon point that was 8000 years old. Desmond Clark identified this and other bone harpoon points found by Don as belonging to the Mesolithic Early Khartoum Tradition—the first time it had been dated. Don and I visited the Tagra pit site twice, once in 1972 and again in 1973. On the first occasion I was clean-shaven; on the second I had regrown my full beard. The Tagra village chief, who had known me when I was bearded in 1964, took me aside and said: 'there was a fellow here last year pretending to be you, but we knew he wasn't! Welcome back again after all these years!' It was now dawning on us that revisiting an old site can often reveal new scientific insights.

Don and I also made our way to the twin granite hills of Jebel Tomat where I had found wild honey ten years earlier. Here Don made an amazing discovery. He noticed that some of the broken pottery scattered on the ground had tiny hollows on the surface. He scraped at some bits of pot with his Swiss Army knife and examined the residue by the light of the sun with his pocket Macpherson's microscope. (A doctor named Macpherson, who had been imprisoned by the Japanese in Changi during WW2, invented this microscope. He needed one small enough to masquerade as a cigarette packet, with enough magnification to be able to detect malaria parasites in the blood and which did not need electricity for light.) Don became quite excited and invited me to look. What we could see were tiny cylinders made up of silica and tiny broken hollow spheres that were black inside. Don immediately recognised that we were looking at the remains of freshwater sponges that had been collected from the swamps of the White Nile, gently broken up and used as temper by prehistoric potters. The silica rods had much the same role in strengthening the pots as do the steel rods used today in reinforced concrete. We were later able to date the pots as being about two thousand years old. This was the very time when the mad bad Emperor Nero had despatched two unfortunate centurions and their troops to find the sources of the Nile. They reached as far as impenetrable swamps from which emerged a granite mountain. This mountain today lies 300 km to the north of the nearest swamps. It was also the time when the Greek historian Diodorus Siculus reported that cattle rustlers used to swoop down from the Red Sea Hills, steal cattle, and vanish with their booty into the swamps of those mountains, where they were safe from reprisals. There are no swamps in those mountains today, but there were back then, as I was to discover during a visit to Erkowit in the Red Sea Hills in January 1973.

Early in 1973 Desmond Clark arrived with his team of three graduate students (Andy Smith, who had been with me at Adrar Bous in 1970, Dan Stiles, who later

Fig. 9.3 House built for Sir Henry Wellcome during excavations at Jebel Moya, 1911

Fig. 9.4 8000 year-old shells of the semi-aquatic edible snail *Pila wernei* on the Shabona Mesolithic site, lower White Nile valley

joined the United Nations Environment Programme in Nairobi, and Ken Williamson, who later worked with us in Ethiopia, where he showed more skill as an entrepreneur than as an archaeologist). We excavated at three localities: Jebel Tomat, which revealed evidence of early sorghum domestication 3000 years ago, Jebel Moya where Sir Henry Wellcome had excavated on a vast scale before WW1 (Fig. 9.3), and the Mesolithic site of Shabona on a sand dune near the White Nile, which had been used as a seasonal camp by small groups of hunter-fisher-gatherers who used to collect *Pila* shells from the swamps nearby some 8000 years ago (Fig. 9.4).

In 1976 some friends from my early days in Sudan with Hunting Technical Services (see Chap. 5), namely Gerald Wickens, a botanist who was now based at Kew Gardens near London, and Martin Adams, who was now HTS Technical Director for Sudan, invited me and three close colleagues to go and assist David Parry who was mapping the very complex soils around Jebel Marra volcano in Darfur Province of western Sudan. Don and I had come from Sydney while Bill Morton and Frances Dakin flew in from Ethiopia (Fig. 9.5). After a short reconnaissance trip up the Blue Nile in the now very battered Land Rover Don and I had shipped over from Sydney for the work with Desmond in 1973, the four of us met up with Dave Parry at his camp in the small town of Nyala. Dave was a genial and very fit Welshman who had played rugby for the Barbarians and had studied soils at Oxford with my Macquarie friend and colleague Ron Paton, an iconoclast if ever there was one! We had a wonderful time deciphering the ever-changing environments and climates in this huge area over a time span of about 500,000 years. Only by doing so could we make sense of the distribution and properties of the soils in this region.

Fig. 9.5 Logging a geological section near Jebel Marra volcano, western Sudan: Bill Morton, Frances Dakin and the author, January 1976

Fig. 9.6 Salt lake inside Deriba volcanic caldera, Jebel Marra, western Sudan, January 1976

Fig. 9.7 Oil palm leaf fossils, Jebel Marra, western Sudan

Jebel Marra rises to an elevation of 3042 m and covers a huge area of 13,000 km^2. Inside the main caldera there are two lakes, one deep and brackish, the other virtually dry and highly saline (Fig. 9.6). Permanent freshwater springs on the mountain have made it attractive for humans well back in prehistory. We found evidence in the form of stone tools of at least episodic human occupation extending back well over half a million years. One spectacular earlier find by Gerald Wickens was of fossil oil palm leaves (Fig. 9.7) embedded in the volcanic tuffs. Gerald thought they were quite young (only a few thousand years) but the associated Early Stone Age stone tools above and beneath the fossil beds indicated an age between about 0.8 and 1.3 million years. The slopes of the mountain were extensively terraced and were still cultivated by the local Fur people native to Darfur. In times past droughts caused occasional conflict between sedentary farmers and nomadic camel herders moving south to graze their animals on the farmers' land. Traditional laws enabled this process to proceed peacefully. If disputes arose, the tribal leaders of both parties settled them swiftly. Unfortunately, the central government has removed the powers of the former leaders and replaced it with heavy-handed central control. The current tragic events in Darfur are in part a result of this unfortunate policy.

For some years now, Don Adamson and I had been collecting and dating freshwater snail shells from the Blue and White Nile and their former channels and flood plains. These shells ranged in age from modern back to about 15,000 years in age and we now had a fair idea of the flood history of these two major tributaries of the main Nile. By now I felt that the time had come to put on record all we had discovered about the geological history of the Nile and of the Sahara into a more permanent form than scattered scientific papers. The result was two books: *The Sahara and the Nile* (1980), which I edited with Hugues Faure, and *A Land between Two Niles* (1982), which I edited with Don.

Several decades later a group of limnologists working in Uganda's Lake Victoria claimed that the present flow regime of the White Nile (and hence the main Nile) was not established until about 7000 years ago, which struck us as nonsense. With my geologist friend Mike Talbot, now at the University of Bergen in Norway, we figured that if we analysed the strontium isotopic composition of the freshwater snail shells we had painstakingly collected, we could determine with confidence just when the modern hydrological regime of the Nile was established. This we did, confirming that the African summer monsoon was strengthened quite abruptly 15,000 ago, leading to renewed overflow from the Ugandan lakes into the upper White Nile. One clear lesson is never jettison samples: new techniques will arise and allow old questions to be answered.

In late December 1982, Desmond had assembled a formidable team of archaeologists and earth scientists in Addis to continue the work we had begun in early 1981 in the Middle Awash valley of the Afar Desert of Ethiopia (see Chap. 14). Politics intervened and we were barred from going to the Afar. Don, Peter Jones (who had worked with Mary Leakey at Olduvai Gorge in Tanzania) and I then decided to fly on to Khartoum and resume our fieldwork along the White Nile. Peter soon discovered that a large cache of elephant tusks retrieved from poachers was now stored at the zoo. It was too good a chance to miss. He carefully analysed the wear patterns on the tusks and quickly determined that the wear marks claimed by some French archaeologists as definite evidence that ancestral humans used tusks for digging were entirely natural!

Leaving Peter to his elephant tusks, Don and I drove down to Esh Shawal where we camped in a disused school building. Our driver was Musa, a former Sudanese army driver. Early one evening, our work now finished, Don and I were sipping tea with the local Block Inspector when our head digger, a burly Fellata called Mustafa, came to tell us that there was unrest in the village and that the men of the village wanted to kill Musa, who was under protective custody in the police station. We made our way there and I spoke quietly to the sergeant, who was keeping the crowd under control with a long whip of hippo hide. It transpired that our foolish and idle young watchman had scared a schoolgirl coming out from evening class. She had screamed. The boy persuaded her to lie and say it was Musa (who was nowhere near, since he was having a quiet drink with a man called Idris). People had come rushing out and the boy had sown his mischief. I cross-questioned the girl, who quickly proved an unreliable witness, and persuaded the police that no punishment of the boy was needed. All through the night the men of the village came up one by one to apologise to Musa. We drove back to Khartoum next day. Don was very shaken by these events and after that he never wanted to go back to the Sudan.

Chapter 10
Wadi Azaouak, Niger (1973–1974)

In one of his stories (*Fullcircle*, 1920, reprinted in 2008 by the Folio Society, London) John Buchan has one of his characters remark that 'to tackle the future you must have a firm grip of the past'. I could not agree more heartily. It has always seemed to me that many (if not all) of our present social, economic, political and environmental woes have their roots and origins in past events. If we could understand these past events more clearly, we would benefit by having a broader, more multi-dimensional perspective with which to look to the present and to the future. That this can be so was brought home to me during some fieldwork which Mike Talbot and I carried out in the semi-arid Wadi Azaouak region of Niger (Fig. 10.1) in December 1973–January 1974.

Mike was working at that time in Ghana and had become increasingly interested in the recent geological history of the southern margins of the Sahara, an interest

M. Williams, *Nile Waters, Saharan Sands*, Springer Biographies,
DOI 10.1007/978-3-319-25445-6_10

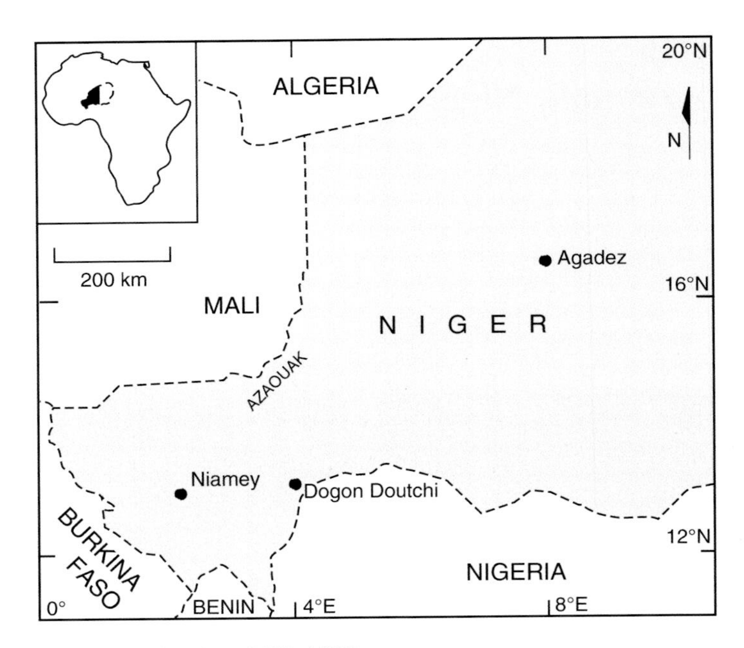

Fig. 10.1 Wadi Azaouak, Niger (1973–1974)

Fig. 10.2 Hotel de l'Aïr, Agadès, Niger, December 1973

that I shared. We hired a couple of camels, purchased goatskin waterbags and basic food supplies from Agadès (Figs. 10.2 and 10.3), and set forth with our Tuareg guide through the acacia grasslands bound for Wadi Azaouak, where early French explorers had reported an abundance of fossil bones and prehistoric stone tools. Soon after our departure a fiery Tuareg lady burst out of the trees, berated our sheepish guide (her husband), telling him in no uncertain terms that the camels were

Fig. 10.3 Mud-brick mosque, Agadès, Niger, December 1973

Fig. 10.4 Morning tea, Azaouak valley, Niger, December 1973. Mike Talbot sitting on *right*

Fig. 10.5 Author and camels, Azaouak valley, Niger, December 1973. *Photo* Mike Talbot

hers and he had no right to hire them out. (The Tuareg had—and may still have—a splendidly matriarchal society, in which the women chose their lovers and future spouses, owned the wells, and owned much of the livestock.) Now deprived of our camels we set forth once more, this time with a couple of donkeys and a grumpy guide.

It was chilly out on the open grassy plains (Fig. 10.4). Mike was wearing a bright orange parka that had been treated with some form of waterproofing that proved attractive to one amorous male camel. The camel blew out a great pink membrane from its mouth and pursued the unfortunate Mike around a solitary acacia tree while I yelled to Mike to punch the camel on the nose. Man and camel ran round the tree for some time until the disgruntled beast took off in search of a more compliant mate.

Some time later we met up with some hardy camel herders, paid off our ineffectual guide, and in due course riding two to a camel we reached a branch of the great Wadi Azaouak drainage system. A quick survey revealed that we could not achieve anything worthwhile in the time available to us because there were no good gully exposures or geological sections and we were not equipped for a major excavation. We headed back towards Agadès through the spiny *cram-cram* grass (*Cenchrus biflorus*), which may well be a useful form of famine food but made walking unpleasant. (We had not brought gaiters.) Mike wanted to check out some gullied dunes he had spotted from the back of the truck in which we had earlier travelled from the capital Niamey to Agadès.

Equipped with fresh camels (Fig. 10.5) and an aristocratic young Tuareg guide and his younger brother, we walked through the dunes and past waterholes from which we replenished our water supplies. The water was green and muddy and I offered the use of my shirt to serves as a filter, an offer promptly rejected. Our guide despatched his younger brother with a small axe to collect some bark from a nearby tree. He added a few bits of this bark to the green sludge in our calabash. Within about a minute the organics had flocculated and settled to the bottom of our container. We then decanted the clear liquid above into the goatskin water bags and also brewed up some tea. The tree with this remarkable water clarifying capacity was none other than *Boscia senegalensis*, a tree I knew well from my time in Sudan, but I had never previously come across this useful property.

In due course we found the gullied dunes Mike had noticed and spent some time mapping the gullies and examining the sediments exposed in the gully walls. These consisted of an alternation of sands washed off the vegetated and presently stable dunes during occasional very heavy rainstorms and buried soils that had developed on the sands once they had been colonised by plants. We found some charcoal in one of the upper layers and once the charcoal had been dated, Mike and I realised that the sediments exposed in the gullies covered a probable time range of about ten thousand years. What had happened during this time was periodic erosion of the vegetated and stable dunes and a type of natural healing process during which plants recolonised the dunes and soils developed. In the 1960s and 1970s it was fashionable in some circles to blame human mismanagement and overgrazing for the erosion of the vegetated dunes along the southern margins of the Sahara. Our work showed that purely natural processes could lead to land degradation and desertification and that in time the landscape would revert back to its vegetated and stable condition.

This is not to say that overgrazing does not occur or that the occasional drought is of no consequence: on the contrary. Soon after our initial arrival in Agadès I was chatting to a Tuareg man about the impact of the drought upon his herds and his extended family. We shared a small pot of tea in his tent. A small girl crawled in and he gave her the tea-leaves to chew. He had lost everything. His camels had died. He had moved south to Agadès to seek help. Out in the market sacks of millet donated by the American government were being sold at exorbitant prices. I bought one and gave it to my host; it would tide them over for three months, he told me. I made some discreet enquiries, since these were intended as gifts to be given freely to those in most need. Apparently unknown to her husband, the wife of Niger's first President (Hamani Diori) was running a black market profiteering racket and was now a very wealthy woman, while her people starved. She was later shot during a military coup in 1974 while he spent the next ten years in prison.

It was on the return trip to Niamey that I saw my first (and, I hope, my last) Gaboon viper. These deadly snakes are quite large in diameter and are incredibly well camouflaged among the leaves and undergrowth in the southern forests. We also saw a traditional lion hunter walking through the market in the small town of Dogon Doutchi. He was clad in a lion-skin cape and carried an enormous bow and quiver of arrows. Everyone moved well out of his way to let him pass and he ignored the bystanders completely. He reminded me of the rare visits of some wandering Hadendowa tribesman into the market at Wad Medani in central Sudan, with similar impact on bystanders. Renowned for their warrior qualities, these fierce tribesmen, led by Osman Digna, were the ones who broke the British square in 1885 at the battle of Tofrek near Suakin on the Red Sea during the early days of the Mahdīya.

My return to Niamey involved some minor drama. I was booked to fly on to Algiers and Paris and so to Ethiopia to meet up with Desmond Clark and his team. On arriving at the airport I found it encircled in barbed wire with machine-gun emplacements at close intervals. A group of armed soldiers ordered me to halt, so I asked to see the officer in charge. The man who came out of his tent hitching up his trousers did not seem quite with it, and was less than amiable at having his rest disturbed. I was escorted away under armed guard. It was some days before things settled down and I was able to leave.

Although the scientific results of our foray into the Azaouak valley were quite minor, our experiences there had provided seeds for reflection. I had seen at first hand the impact of a major drought upon both peasant farmers and nomadic pastoralists. I had also witnessed what may be called 'natural desertification' in the form of eroded dunes and gullied fans. By now I was starting to ask myself what were the ultimate causes of drought and could they be foreseen and monitored with a view to providing timely forecasts to local communities about future droughts. Finally, I was becoming increasingly drawn into the complex and thorny problem of the relative role of human activities and climatic fluctuations in causing degradation in the dry lands that cover over a third of the land area of the earth.

Chapter 11
Petra and Wadi Rum, Jordan (1975)

The Quaternary Period covers the last 2.6 million years of geological time and was a period of frequent and rapid changes in climate in every region on earth. In particular, it was a time when great ice caps waxed and waned, sea levels rose and

M. Williams, *Nile Waters, Saharan Sands*, Springer Biographies,
DOI 10.1007/978-3-319-25445-6_11

fell, deserts expanded and contracted, lakes changed from fresh to saline to fresh, and plants, animals and humans migrated, adapted or became extinct. For me it was the most exciting interval in geology, not only because the evidence of past environmental changes was often so well preserved, especially in the drier regions, but also because it was the time when we became fully human.

Just outside Paris, in the suburb of Meudon-Bellevue, there was a first rate research institute devoted to studying this tumultuous interval of geological time. This was the *Laboratoire de Géologie du Quaternaire*, funded by the French *Centre National de la Recherche Scientifique* (CNRS). The foundation director of this Quaternary geology research institute was Mademoiselle Henriette Alimen, who had carried out pioneering work into the prehistoric archaeology and Quaternary geology of the great Saoura river basin and the great western erg (sand sea) in the Algerian Sahara in the 1960s.

Her successor was Professor Hugues Faure, with whom I later edited a contributed volume called *The Sahara and the Nile* (1980), and whom I had first met in Ethiopia in December 1971. We shared a common interest in the geologically recent climatic fluctuations in the Sahara. Hugues very kindly agreed to my request to spend a sabbatical year in his research institute at Meudon-Bellevue. There were some brilliant researchers at Bellevue. These included Raymonde Bonnefille, who had carried out detailed studies of the vegetation history of Ethiopia based on analysis of fossil pollen grains preserved in lake and swamp deposits; Maurice Taieb, who had led the 1971 excursion to the Afar (see Chap. 8) and who was completing his studies of the entire Awash basin in Ethiopia; Michel Icole whom I had met in Niamey (Niger) two years earlier, and Nicole Petit-Maire, who was to lead future expeditions into remote parts of Libya and Mali.

I shared a laboratory with Françoise Gasse, who was writing up her work on the lakes of the Afar. Trained as a geologist, Françoise became one of the world's leading authorities on the use of diatoms (tiny algae with a skeleton made of silica) to reconstruct past changes in lake water chemistry, temperature and depth. Using this type of evidence it is possible to reconstruct past changes in climate with some precision. Françoise and I became life-long friends and colleagues. She and Raymonde Bonnefille were later to work with Desmond Clark's geo-archaeological team at Gadeb in Ethiopia.

While in Paris, I met up once more with Professor Pierre Rognon who was head of physical geography at the Université Pierre et Marie Curie at Jussieu. He and I worked together on a long review article comparing the climatic histories of the Australian and North African deserts and their margins. One reason why I have always been attracted to working in deserts is that the evidence of past environmental changes is often very well preserved thanks to the aridity prevalent in the desert. In a sense, deserts function as vast environmental museums, so that prehistoric rock paintings, fossil bones, ancient river and lake sediments, buried soils and archaeological middens have suffered minimal damage from wind, water or ice.

Out of the blue the Polish hydro-biologist Dr Julian Rzóska asked me to read and correct the draft geological chapters in his edited manuscript book on the biology of the Nile. We met in the Bishop's House at the Palais des Irlandais in Paris and talked into the night. It was wonderful to meet up with another Nile enthusiast who had also worked extensively in the Sudan. In the Preface to his own contributed volume on the Nile, Henri Dumont gives an amusing account of this encounter.

At the end of that year Pierre, Françoise and I carried out work on some of the older sediments in the Afar, while I showed them around the site of the ancient lake at Gadeb up in the southeast highlands of Ethiopia. While we were down in the Afar Desert, in what was then the French territory of TFAI surrounding the strategically important port city of Djibouti, we had some instructive experiences. (TFAI stands for *Territoire Français des Afars et des Issas*). One evening we were chatting quietly in our camp about fifty metres from a dirt road through the mountains when we saw a convoy of several hundred camels loaded with long rectangular boxes being led along silently by well-armed men. We kept out of sight, relieved that our campfire was long dead. Next day we reported what we had seen to the officer in charge of a French Foreign Legion fort near Dikhil. He listened alertly, muttered something about gun smuggling and indicated that he and his men would soon apprehend the caravan.

On another occasion the French commander of a fort garrisoned by local soldiers had invited us to join him for lunch. We arrived punctually to be informed that the commander would be with us shortly, once he had returned from his twenty kilometres run with his pet cheetah. Lieutenant François de Barbeyrac was a delightful and exceedingly fit individual from Gascony in the southwest of France.

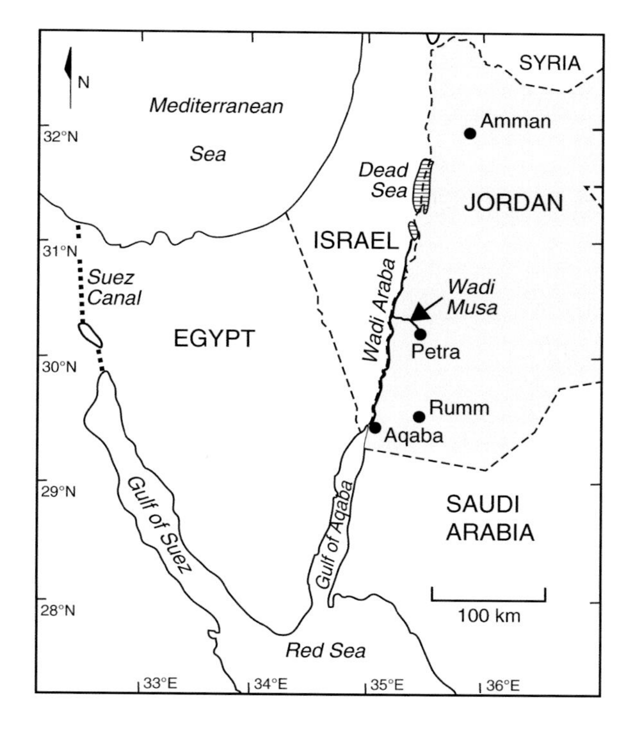

Fig. 11.1 Petra and Wadi Rumm, Jordan

Fig. 11.2 **a–c** Petra, Jordan, 1975

The Gascon people have a reputation for valour and eccentricity, and he was no exception. His dream was to run across Australia in summer from Darwin to Adelaide—a mere two thousand kilometres. When, a little later at army headquarters in Djibouti, his General asked me if the Lieutenant de Barbeyrac would succeed in this quixotic endeavour, I straightened my shoulders and replied without hesitation '*Oui, mon Général*!' This pleased him greatly and he immediately offered

Fig. 11.3 **a**, **b** Wadi Rumm, Jordan

us the use of a French army helicopter to help us get to some of our less accessible sites. De Barbeyrac achieved his run across the Australian desert in summer—an amazing feat—and then went on to ski in New Zealand. He later offered to ship several tonnes of elephant fossils to me in Sydney if I wanted. I declined, very firmly, and indicated that the natural history museum in Paris was the most appropriate place.

I had learned earlier in the year that my father had cancer. While he was still able to travel we arranged a family holiday at the seaside town of Aqaba in Jordan (Fig. 11.1). Dad especially enjoyed his visit to the ancient city of Petra (Fig. 11.2a–c).

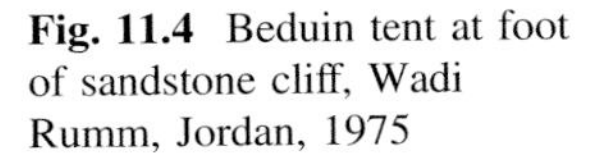

Fig. 11.4 Beduin tent at foot of sandstone cliff, Wadi Rumm, Jordan, 1975

Petra (Fig. 11.2c–d), described by the young Oxford poet John Burgon in 1845 (who, rather ironically, had never been there) as the 'rose-red city half as old as time'. Petra had been carved by Nabatean masons from the multi-coloured sandstones of the Jordanian desert, not far from Wadi Musa, some two thousand years ago. It was at the site of the existing spring called Ain Musa in Wadi Musa that Moses struck the rock to obtain water. The local Beduin who informed me of this spoke quite matter-of-factly, as if it had only happened yesterday. Not far from Petra is Wadi Rumm, with its magnificent sandstone cliffs and gorges (Figs. 11.3a, b and 11.4). Halfway up one cliff is a spring at the contact between the overlying sandstones and the granites beneath, with a large and presumably Nabatean sandstone lower grindstone nearby. The Nabatean farmers were masters at harvesting runoff and so were able to grow viable crops even in the arid southern Negev Desert of Israel. I suspect that the climate was somewhat less arid then.

The sandstone cliffs of Rumm greatly impressed Colonel T.E. Lawrence ('Lawrence of Arabia') while serving with the Arab irregular forces on their way to attack the Turkish garrison at Aqaba on an inlet of the Red Sea. In *Seven Pillars of*

Wisdom (1935) Lawrence especially appreciated the chance to bathe in a small pool on a ledge in the sandstone, and was enjoying the cool water when there suddenly appeared a 'grey-bearded, ragged man, with a hewn face of great power and weariness' who used to wander in these parts, muttering to himself, fed by hospitable passers-by. Lawrence referred to him as 'the old man of Rumm'. He also noticed Nabatean inscriptions on the cliff face, together with 'Arab scratches, including tribe-marks, some of which were witnesses of forgotten migrations'.

Chapter 12
Algeria and Tunisia (1979)

During late 1978 to mid 1979, I returned to Paris as a Visiting Professor in Pierre Rognon's department at the Université Pierre et Marie Curie at Jussieu in the heart of that great city. It was an excellent opportunity to make new friends and catch up with old ones. The French tradition of a leisurely lunch allowed ample time for discussion and I came to value yet again the willingness of my Gallic colleagues to hurl ideas into the air and throw them around with gusto. I came to value this "flow of soul and feast of reason", not least because I had ideas I wanted to test on others. By now I was putting the finishing touches to the big volume on The *Sahara and the Nile*, which I had first conceived during my 1975 visit to Paris, and was checking the draft chapters of the slimmer volume called *A Land between Two Niles*.

Soon after my arrival in Paris I went to visit Théodore Monod in his office in the Natural History Museum. Monod was the greatest Saharan scientific explorer of his

M. Williams, *Nile Waters, Saharan Sands*, Springer Biographies,
DOI 10.1007/978-3-319-25445-6_12

day and a delightful man with a great sense of humour. He was a real polymath, spoke and read many languages, and was a stickler for scientific accuracy. He agreed to write a foreword for *The Sahara and the Nile*, which I was preparing for publication at that time, and I also explained the circumstances of Bill Morton's death in Ethiopia two years earlier. Bill was a keen caver and had guided Monod on a trip to the Sof Omar cave, the longest cave system in Ethiopia. Here Monod had discovered a blind, translucent cave louse new to science. He named it *Scotobinus mortoni* in Bill's honour and Bill was tickled pink to have such a creature named after him. A few weeks later Monod invited my friend and colleague Dr Nicole Petit-Maire and me to join him and his wife for lunch in his apartment on an island in the Seine. The food was frugal and served with water, but the conversation was wide-ranging and culminated in a discussion of historic shipwrecks off the coast of Mauritania near the Banc d'Arguin (which I was to visit in 2004). Lunch over, Monod handed me my overcoat (it was winter) and with a brisk '*Allons, au travail*' ('Come on, off to work!') set forth briskly for work once more. Nicole, meanwhile, greatly in need of some reviving wine, persuaded me to slip into a nearby café.

On another occasion I had lunch with Professor Jean Dresch, who knew the Sahara well, and who had come to see the Australian desert in 1973, after a long spell of rain. On seeing this pathetic version of a desert he was moved to ask: 'The Australian desert! Where is the Australian desert? This is the Sahara of Neolithic

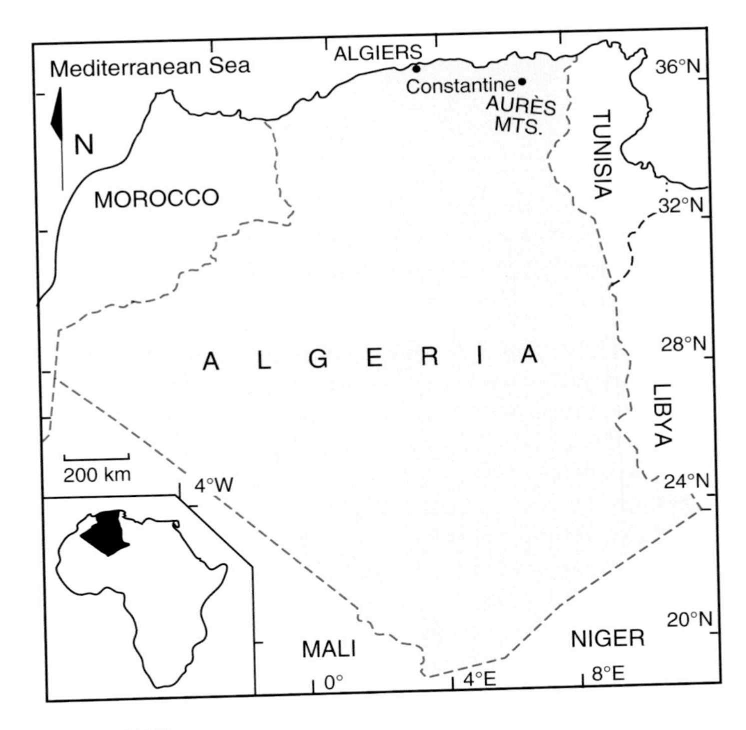

Fig. 12.1 Algeria (1979)

times' (when it was much less arid than today). Over lunch I was rash enough to mention that we had some very good wines in Australia. He was quick to retort: 'Some good wines, yes; very good wines, no!' Some years later, when savouring a 1958 Chateau Margaux, I had to admit that Dresch might well have been right!

As a welcome break from delivering three-hour lectures to an attentive fourth-year class of university students, I eagerly accepted two invitations to take part in fieldwork in Tunisia and Algeria. The fieldwork in Algeria was at the invitation of Professor Roger Coque's Algerian doctoral student Mohamed Tahir Ben Azzouz, with whom I became good friends. Mohamed was a lecturer at the University of Constantine (Fig. 12.1) and was studying the high plains of eastern Algeria. It was a land of salt lakes, gypsum dunes and gypsum sand plains covered in saltbush with the rugged Aurès Mountains to the south. We worked hard digging, boring holes with my soil auger, describing, and sampling. Late one afternoon we returned to the small shed where we used to leave our few belongings to find that our waterbag had been slashed with a sharp blade and our drinking water was gone. The owner of the shed, a hospitable old man, was embarrassed and fearful. He had also trapped some rabbits. I suggested that he cook up a rabbit stew and invite those who seemed to object to our presence in those parts to share a meal with us, so that we could explain why we were there.

They came: a small group of hard men who had fought the French for years in the battle for independence. We exchanged stories. I told them of one episode when a recently shot buffalo was blocking a steep and narrow track leading up to a tin mine on top of a hill in Northern Territory, Australia. My wild Irish miner friends the Toohey brothers were operating this mine. John Toohey had gone for a walk and had narrowly escaped a charging buffalo. (The Northern Territory buffaloes are water buffaloes brought in from Timor a century earlier and are now quite wild). Paddy protested that only having one leg he could not run away from an angry male buffalo. They took out their trusty Lee-Enfield .303 rifle and downed the buffalo as it charged. A big snag was that the truck from the Mount Wells battery was due any moment to collect a load of their tin ore for crushing, and would come round a blind corner straight into a ton of dead buffalo. With great glee they described what they did. 'So, we stuck six plugs of dynamite up one end and six plugs up another'. They cannot have been thinking too clearly because they certainly removed any trace of buffalo from the landscape but also made a huge crater in the track...

Luckily, this grotesque tale seemed to appeal to the macabre sense of humour of the Algerian former partisans. They relaxed visibly (the rabbit stew helped) and began to tell their own stories. One was how they had evaded capture from a company of French paratroopers with fixed bayonets who suspected that a small group of about a dozen partisans were hiding out on the salt bush plains that stretched out in front of us as we ate. The parent soil in this region is wind-blown gypsum, which is very cohesive, easy to carve and not prone to collapse if dug into. The Algerian partisans had previously dug a spherical underground bunker with only a narrow entrance at the top, easily camouflaged. A dozen armed men with food and water could lie hidden in such a shelter for days if necessary, without the French ever suspecting.

Fig. 12.2 Tunisia (1979)

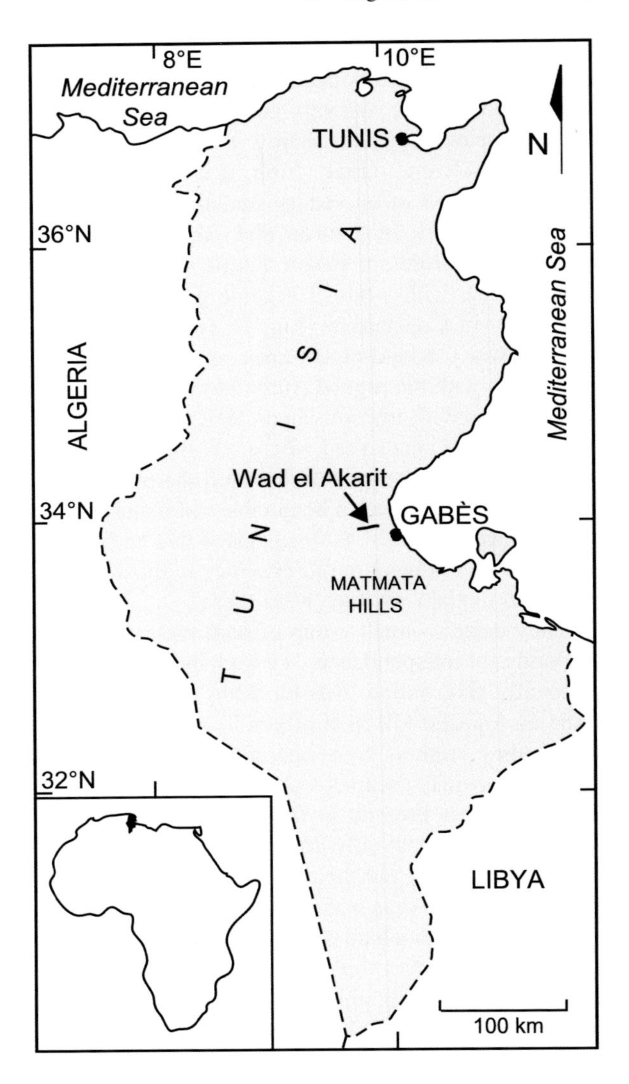

I was glad when we finished our work in this area. We returned to Constantine where we enjoyed the luxury of the local *hammam* (the Turkish and Arab version of the old Roman hot baths) with a powerful masseur scraping away the accumulated gypsum dust from our pores. Thoroughly relaxed, I went to give a few lectures next morning at the University. Crossing the bridge over a deep and relatively narrow limestone gorge near the University, I suddenly realised that I had seen that very place before. My father had helped build a temporary bridge across the ravine during his time with the Royal Engineers in WW2, and I had seen his photos.

The second bit of fieldwork was in Tunisia (Fig. 12.2) and was organised by Pierre Rognon. It proved a great deal more luxurious than the Spartan conditions of the Algerian fieldwork! During the first leg of our work we stayed at Gabès and tried to make sense of the complex depositional history exposed in the vertical banks of the Wad el Akarit. This long, very straight and wide channel gives the impression of having been cut by a gigantic saw and connects the arid inland salt-flats to an equally arid coast. It proved a major defensive asset for Field-Marshall Rommel and the Afrika Korps in their final retreat from northern Libya in 1943.

Our party consisted of Pierre Rognon, Jean Riser, Geneviève Coudé, Alain Lévy, Jean-Louis Ballais and the brilliant isotope geochemist and all round geologist Jean-Charles Fontes, who became an esteemed friend and later stayed with us in our home in Mornington, Victoria, Australia. Geneviève was studying for her doctorate supervised by Pierre Rognon and later became an authority on land degradation (desertification) and land management in northwest Africa. Alain was trying to determine whether there had been incursions of the sea and so was on the lookout for marine shells and microscopic marine fossils. Jean Riser and Jean-Louis Ballais had wide experience of alluvial deposits in northwest Africa, so we had a very good team.

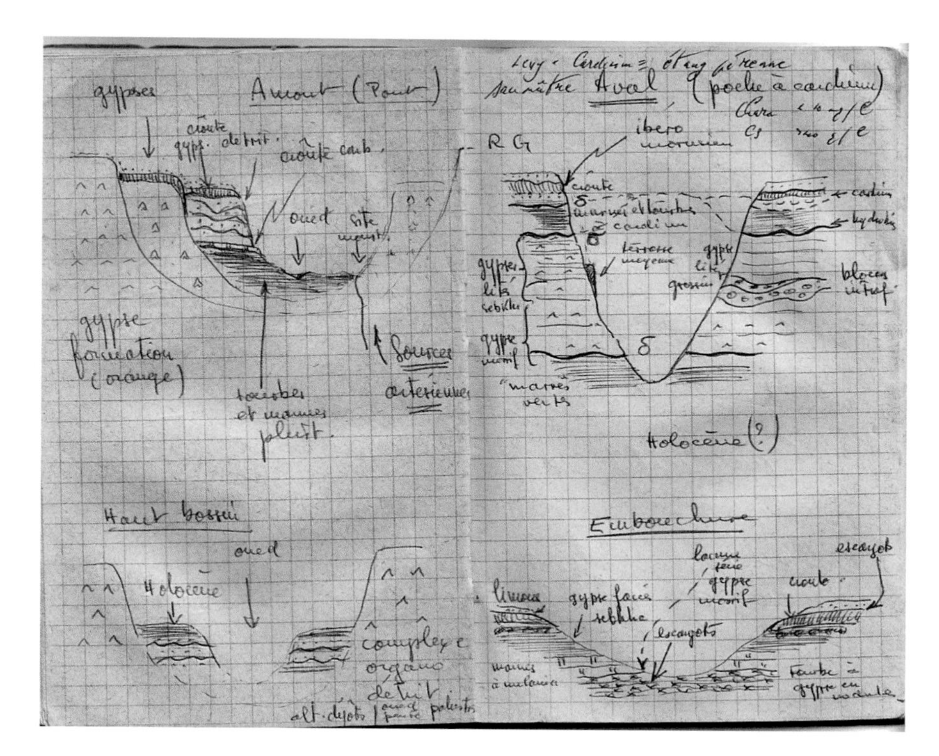

Fig. 12.3 Geological sections of Wad el Akarit sketched in my note book during breakfast on 15 April 1979 by Professor Jean Charles Fontes. *Amont (pont)* is 'upstream (bridge)'. *Haut bassin* is 'upper catchment'. *Aval* is 'downstream'. *Embouchure* is 'outlet' or 'mouth'

Fig. 12.4 **a**, **b** Roman ruins, Matmata Hills, Tunisia

One morning over breakfast Jean-Charles reached over for my field notebook and rapidly sketched in the highly complex stratigraphy exposed in the banks of the Wad el Akarit (Fig. 12.3). When I subsequently showed the sketch to Jean Riser, who was no slouch at this type of work, he remarked with admiration and a tone of awe: '*L'oeil du maître*!' ('The eye of the master').

Having come up with an acceptable interpretation of the Wad el Akarit sediments, Jean-Charles Fontes, Alain Lévy and Jean Riser returned to their teaching and research duties in France. The rest of our party moved inland to the high limestone country of the Matmata Hills. Three things immediately impressed me about this region. First, scattered through the landscape there were innumerable Roman ruins. Two thousand years earlier the Romans had occupied this region and obtained an abundance of wheat, barley and olive oil from the small valleys. They used an ingenious technique to obtain cultivable land—building stone dams across the valley floor to trap silt while allowing runoff to flow through the porous dam wall (Fig. 12.4a, b). In his classic book *The Mediterranean Valleys* (1969) my good friend Claudio Vita-Finzi provides a scholarly and witty description of these dams.

The streams have cut down many metres since Roman times, leaving the former dams high and dry. However, what struck me next was that the present-day farmers in this area were using identical methods to the Romans, building walls of stone to trap river silt along the valley floors and hillsides (Fig. 12.5a, b). The largest of these patches of sediment supported quite large groves of date palms and olive trees; the smallest might only have a couple of trees, but could be expanded over time.

The obvious question to ask was whence came the silt. The limestone hills were often pretty bare and barren, with a few shrubs and bushes providing a meagre feed for the flocks of hardy goats (Fig. 12.6). However, in sheltered patches in the landscape there were deep and often extensive deposits of yellow-brown or red-brown silt. This material was in fact wind-blown desert dust, and had been

Fig. 12.5 **a**, **b** Palm trees and olive tree growing in silt trapped behind stone-walls, Matmata Hills, Tunisia

deposited across the landscape during windier, dustier times some twenty thousand years ago. In places the desert dust or *loess* was sufficiently thick to allow caves to be fashioned in it to be used as dwellings. We slept one night in one such cave dwelling. A decade later, during a visit to the great Loess Plateau of central China, I was to see similar cave dwellings carved in the very thick loess deposits there, which attained a remarkable thickness of 200–300 m in many places, and reflected well over a million years of desert dust deposition.

The third thing to attract my attention was the peculiar nature of the ephemeral stream channels in this arid environment. The modern streams flowed only occasionally and when they did they carried a coarse load of limestone cobbles, very

Fig. 12.6 Goats grazing on barren limestone hills, Tunisia

Fig. 12.7 River bank sediments derived in part from reworked desert dust, Matmata Hills, Tunisia

much as one might expect of any stream in this type of desert environment. However, flanking these ephemeral stream channels there were vertical banks 10–15 m high made up of very fine sediment (Fig. 12.7). The sediment had originated as wind-blown dust laid down across the landscape and later washed down into the valley bottom before the current arid climate had set in. On the surface of these steep silty cliffs we found prehistoric sites with an age of about 20,000–15,000 years. The present phase of river channel incision must have begun soon after fifteen thousand years ago, when the amount of desert dust accumulating in the hills and valleys had been drastically curtailed, for whatever reason. I was puzzled by this and resolved to pursue the matter later. At the time I did not realise that the answer to the puzzle resided in the arid valleys of Namibia, Sinai and the Flinders Ranges of South Australia.

Back in Paris, I resumed teaching and writing. One day I was walking beneath some ornamental eucalyptus trees when it began to rain. The smell of damp eucalyptus leaves took me back in spirit to my adopted homeland of Australia. I decided that the time had come to leave the fair city of Paris and move southwards. Besides, there were some interesting conferences and field excursions coming up shortly in South Africa—just a hop away from the land of gum trees…

Chapter 13
Son and Belan Valleys, India (1980, 1982, 2005)

By a curious quirk of human perception, we tend to see what we expect to see and so we often fail to notice something we are not expecting to see. Let me give an example. On the afternoon of February 5, 1980, geology Honours student Keith Royce and I were walking along the left bank of the Son River in semi-arid north-central India (Figs. 13.1, 13.2, 13.3 and 13.4). We were part of an international team of archaeologists directed by Professor G.R. Sharma from the University of Allahabad in India and Professor Desmond Clark from the University of California, Berkeley. Our task was to define and map Quaternary sedimentary formations associated with prehistoric stone artefacts ranging in age from Lower Palaeolithic through Middle and Upper Palaeolithic to Mesolithic and Neolithic. In

M. Williams, *Nile Waters, Saharan Sands*, Springer Biographies,
DOI 10.1007/978-3-319-25445-6_13

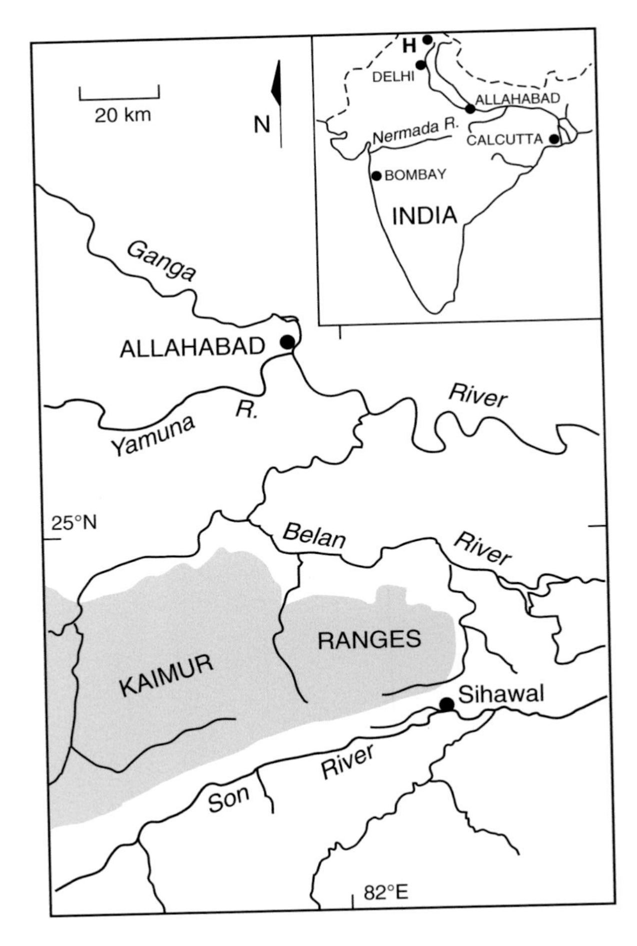

Fig. 13.1 Son and Belan Valleys, north-central India (1980, 1982, 2005). The letter H in the *top of the map* denotes the Haritalyangar area in the Himalayan foothills of northern India, where we worked in 2001 and 2003

this we thought that we had been reasonably successful until we were confronted with an unusual deposit near the foot of a 35 m high bank section close to the junction of the Son and Rehi rivers. We had seen nothing like it before. I remarked casually to Keith that if this had been Ethiopia, I would have called it a volcanic ash or, at least, a diatomaceous ash (Fig. 13.5), but no Quaternary ash had ever been recorded anywhere in India. I decided to sample it at close intervals.

The deposit was indeed a volcanic ash and proved to have originated from a huge eruption of Toba volcano in Sumatra some 74,000 years ago. Every explosive volcanic eruption is unique, and the ash from such eruptions is also unique, much as human ear shapes and fingerprints are unique to one individual person. Using appropriate forms of geochemical analysis it is possible to fingerprint ash from a particular eruption. This allows us to say that even widely scattered ash deposits such as the Toba ash deposits in India all belong to the one eruption.

Fig. 13.2 Temple near Patpara, Son valley, north-central India, 1980. *Photo* Desmond Clark

Fig. 13.3 Reclaimed gullied land, middle Son Valley, north-central India, 1980

Once put on record, the ash prompted a widespread search by Indian geologists for other exposures. It soon became evident that all of peninsular India had been covered in a mantle of ash 10–15 cm thick. Louis Pasteur once famously remarked that chance favours the prepared mind. He was of course right. Having worked in volcanic regions already, I was better prepared than colleagues who had never seen volcanic ash. The impact of this event remains controversial, but as far as India is concerned, we now have a marker that enables us to examine environmental changes before and after deposition of the 74,000 years old Toba volcanic ash. The potential of this chance discovery has yet to be fully exploited but it has already

Fig. 13.4 Author and baby monkey, Son valley, north-central India, 1980. *Photo* Keith Royce

triggered a resurgence of archaeological activity and associated environmental research at a number of sites in India. It has also sparked a vigorous and often polarised debate. I will say more about this later.

Professor Sharma had learned his trade working with Sir Mortimer Wheeler, who had directed the great excavations of the ancient cities of Harappa and Mohenjo-Daro in the Indus valley. Sharma himself later directed the excavation of Kaushambi, a city that flourished on the banks of the Yamuna River some 3000 years ago. Sharma was very much the benevolent autocrat and directed all the digging operations with minimal fuss and great skill. He was a very good all-round archaeologist with a particular interest in the origins of agriculture in this part of north-central India. He met Desmond at a conference and invited him to join his team excavating prehistoric sites along the Son and Belan valleys in the Vindhyan Hills south of Allahabad. Desmond agreed, intrigued by the opportunity to investigate a very long record of human occupation in this part of Asia.

Soon after our arrival, Professor Sharma and his archaeological team took us on a tour of the main prehistoric sites in both valleys. One immediate problem we perceived was the lack of an independent stratigraphic control for the prehistoric sites, which is where I could contribute. What they had done, understandably but inappropriately, was to devise a scheme in which certain river gravels were claimed to be associated with a particular set of stone tool assemblages, and vice versa. So, for instance, they considered that 'Gravel One' was 'Lower Palaeolithic', 'Gravel Two' was 'Middle Palaeolithic', and 'Gravel Three' was 'Upper Palaeolithic'.

But nothing is ever that simple. For a start, there were many different gravel units, not just three. In addition, and far more important, river gravels by their very nature have been transported from somewhere upstream, as have any stone tools found within them. The tools are therefore not in what we can consider 'primary context'. The river gravels could therefore contain stone tools of quite different

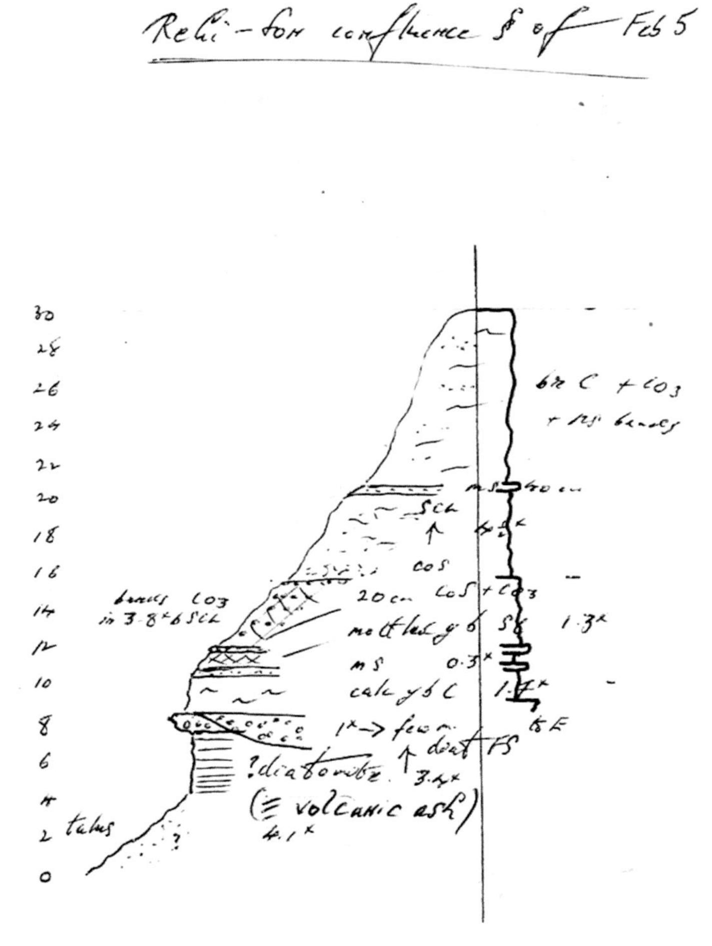

Fig. 13.5 Field sketch section of the volcanic ash we discovered on February 5, 1980. Later work revealed that this ash came from a super-eruption of Toba volcano in Sumatra 74,000 years ago

ages. Our job was therefore to start from scratch and locate undisturbed archaeological sites away from the river. While doing this we also needed to identify depositional units that were widespread along the valleys and to place these into a time sequence that was established by radiocarbon and other forms of dating, independent of the stone tool assemblages. It was quite a challenge and took two full seasons to provide a reasonably solid stratigraphic framework and to establish a reasonably reliable alluvial history. I say 'reasonably' because any field mapping is a work in progress, to be tested against future mapping and new techniques of analysis. Each evening during the first season Keith Royce and I would use chalk and blackboard to give an illustrated account of our progress and explain the elements of river behaviour in particular and Quaternary geology in general.

There were moments of humour, high drama, and unexpected discovery. Early in the first season we passed a large elephant tethered to a tall tree in the village of Sihawal not far from our campsite. B.B. Misra, a tireless walker and fine archaeologist, remarked casually: 'Elephant very fond of eating people!' At least, that's what it sounded like to me. I expostulated: 'Come on BB, pull the other leg!' He insisted, and of course was referring to the *peepul* tree (*Ficus religiosa*).

One rainy night a group of dacoits (brigands) attacked a nearby village store. The courageous daughter of the shopkeeper hurled a spear at the dacoit chief as he was leaving, and struck him in the buttock. He dropped his booty and fled. The police found him later crouching in a ditch. He claimed to have been attacked by a savage cow. All this took place close to our tents so it was a noisy night, with villagers holding flaming brands of wood aloft as they searched for more of the dacoit band.

On another occasion we were excavating a peculiar structure emerging from a layer of alluvial clay we knew to be about 8000 years old. There were distinct tracks of Sambur deer on the surface we were excavating (Fig. 13.6). Some of the local Bega villagers came over and said that we must take great care of the structure, because it was one of their shrines. It was certainly very similar to ones they built today. Here was an example of a practice that had persisted through the ages. The Bega (Fig. 13.7) were a tribal people who still practised hunting with bows and arrows. We accompanied them up onto the sandstone plateau of the Kaimur Ranges (Fig. 13.8) north of the Son valley, where they showed us prehistoric sandstone rock shelters with paintings of rhinos and tigers being caught in pit traps (Fig. 13.9). Two of the Bega men demonstrated how to make fire by rapidly rotating a stick set

Fig. 13.6 Excavated tracks of a prehistoric Sambur deer, Son valley, India, 1982. *Photo* Desmond Clark

Fig. 13.7 The oldest Bega, Son valley, India, 1980. *Photo* Desmond Clark

Fig. 13.8 Kaimur Ranges, north-central India, 1980

vertically into a shallow wooden base covered with grass kindling. They used their hands to rotate the vertical stick and took turns doing so. Flames ensued within 2 min (Fig. 13.10). Later, using an elephant for transport, we excavated one of the rock shelters, which turned out to be Mesolithic in age, and to have been occupied from time to time by the nomadic hunter-gathers who roamed between plain and plateau before the advent of agriculture in this region.

From my point of view the most interesting discovery of all was the presence of relatively pure volcanic ash derived from the eruption of Toba volcano in Sumatra 74,000 years ago. The 1883 eruption of Krakatoa volcano, located between Java

Fig. 13.9 Rock painting of tiger and pit trap, Kaimur Ranges. *Photo* Desmond Clark

Fig. 13.10 Bega tribal men making fire by twirling sticks, Kaimur Ranges, north-central India, 1980

and Sumatra, killed over 40,000 people, and produced about 20 km^3 of ash and associated volcanic rock debris. The Toba eruption, on the other hand, produced at least 2500–3000 km^3 of volcanic ejecta, of which at least 800–1000 km^3 was ash, and has been called a 'super-eruption'. It was probably the largest eruption of the past million years. That being the case, what was its impact?

Trying to determine the environmental impact of a prehistoric volcanic eruption makes for some exciting detective work but is far from easy. With my American colleague Stan Ambrose and my Indian colleagues Jaganath Pal, Umesh

Chattopadhyaya and Parth Chauhan, we returned to the Son valley in 2005 and collected samples of calcium carbonate (limestone) nodules that had formed in soils lying beneath and above the Toba ash at two sites on either side of the present river. Stan, Parth and I also sampled Toba ash at sites in the Narmada valley situated 400 km to the west. (Stan had the misfortune to be attacked and thrown by an irate cow while I was entangled in dense vegetation at the bottom of a deep gully. Happily, the sharp ends of the horns had been removed.) By analysing the carbon isotopes in these nodules we could work out the type of vegetation growing in the region before and after the eruption. We found that forest had been widespread in this area before the eruption, while open woodland and grassland had been widespread after it.

As an additional check, my Dutch colleague, Sander van der Kaars, examined the pollen grains in sediments obtained from a marine core collected from the Bay of Bengal. This sediment core contained Toba ash, so we could see if the pollen grains showed any change in the terrestrial vegetation before and after the eruption. Our joint endeavours suggested a short interval of fairly intense cooling followed by many centuries of below average rainfall in this region.

Other workers have argued that the Toba eruption had no discernible impact on global or regional climate, so that the debate has, unfortunately, now become polarised into two camps: those who argue in favour of an impact, and those who deny that there was any noticeable impact. My own view is that the present debate is unproductive because the evidence put forward is not yet dated with adequate precision. We need to seek evidence with a much finer time resolution, preferably annual, which means working with ice cores, with cave *speleothems* (stalagmites) and with annually layered lake sediments of appropriate age. Whether they like it or not, I am by now pretty convinced that archaeologists, when working strictly as archaeologists, will mostly remain consumers but not producers of useful information about past environments and climates!

It would be remiss not to mention some further work we carried out high in the pine forests of the Himalayan foothills. In 2001 Australian palaeoanthropologist David Cameron invited my good friend Brad Pillans from the Australian National University, his New Zealand geology colleague Brent Alloway and me to work with him and his Indian colleague Rajeev Patnaik on dating the Siwalik sediments in the Himalayan foothills near Haritalyangar. These deposits ranged in age from about twenty to five million years and contained very rare fossils of the Miocene hominids Indopithecus and Sivapithecus, but their age was still in doubt. Brad was an expert at detecting changes in the polarity of the earth's magnetic field, which could be used to work out the age of the fossil-bearing sediments. Rajeev was a genius at finding fossils and an excellent colleague with whom to work. 2001 was the year of the January 26 Gujarat earthquake which occurred on Republic Day and caused massive loss of life. I was in Chandigarh when it took place and felt it clearly. We resumed work near Haritalyangar in 2003 with a small team consisting of Brad, Rajeev, Frances, my PhD student Peter Glasby who had been born and brought up in India, and myself. The work proved successful and we published a full account in the 2005 *Journal of Human Evolution*.

Chapter 14
Afar Hominids, Ethiopia (1981)

Early in 1981, with the security situation in Ethiopia more settled, we were at last able to work out in the Afar or, more precisely, in the middle Awash valley located between the great eastern escarpment of the Ethiopian highlands to the west and the volcanic desert of the Afar to the east (Fig. 14.1). I had visited this area in December 1971 when Maurice Taieb had led a field excursion there to show us some of his fossil sites (see Chap. 8). Desmond Clark and I had worked along the southern margin of the Afar in 1974 and 1975 (Figs. 14.2, 14.3 and 14.4), on sites no older than Middle Stone Age. Here was a chance to work on some of the oldest Stone Age sites in Africa as well as on fossils of Pliocene age (5.3–2.6 million years), which might, if we were very lucky, include hominids. (Hominids are closer to humans than chimpanzees and so may be considered as being our very early ancestors.)

M. Williams, *Nile Waters, Saharan Sands*, Springer Biographies,
DOI 10.1007/978-3-319-25445-6_14

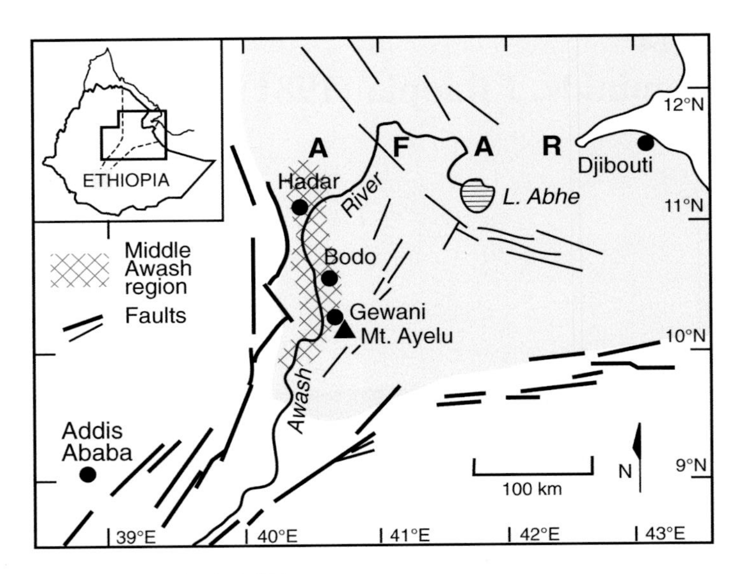

Fig. 14.1 Afar Desert, Ethiopia (1981)

Fig. 14.2 Afar women filling goatskin waterbags, Afar Desert, Ethiopia, 1975

Desmond had assembled a very good team. Tim White would deal with the vertebrate fossils, including any hominids we might unearth. Jack Harris aided by Hiro Kurashina (who had worked with us at Gadeb and K'one), Rob Blumenshine and Carol Sussman (who had been with us in the Son valley in India the year before) would help in excavating Early Stone Age sites and associated animal remains, and Bob Walter would sample the volcanic ash beds for dating. Getaneh Assefa (who had also worked at Gadeb) joined us for a spell, while Mike Tesfaye

Fig. 14.3 Afar mother and daughter with edible lily bulbs collected from a waterhole in the Afar Desert, Ethiopia, 1975. These are used in times of drought. The black seeds inside are ground into flour and made into porridge

Fig. 14.4 Afar children with Ayelu volcano in the background

Fig. 14.5 Collecting water from the Awash River, Afar Desert, Ethiopia, 1981. Author is on the *extreme left*. *Photo* Tim White

was my tough and cheerful field assistant. Desmond's wife Betty saw to what she called the 'commisariat', making sure we were fit and properly fed.

We cut ourselves a track from Gewani through the sometimes quite dense thorn scrub as far as our camp at a place called Bodo by the local Afar cattle herders. We were within a National Park so game was abundant, and the Afar herdsmen grazed their cattle without interfering with the wild animals. Lions, hyenas, jackals and cheetah were quite common, as were great herds of sable-coloured oryx with their long, curved scimitar-like horns. At first we obtained our drinking water by digging into the bed of a dry channel. As the season advanced this supply dwindled and became brackish. After that we went down to the river, waded into the brown turbid water and formed a human bucket chain to fill our water barrels, keeping an eye out for crocodiles or hippos (Fig. 14.5).

One afternoon after my day's work logging and sampling geological sections, I was invited to an Afar camp by the river to tend to a baby who had fallen into the fire and burned his bottom. While I was seeing to the baby, watched anxiously by his young mother, the *balabat* (regional chief) of the local Afar tribesmen came over with his warriors. They were well armed, truculent, and all wore their customary large castrating knives across their waists. After an exchange of greetings, the *balabat* came straight to the point. The young men were angry because our Land Cruisers were disturbing their cattle. However, for the princely sum of ten cents per person per day, he intimated that he could guarantee our safety. I thought this fair and suggested he come to our camp and meet Desmond, known in Amharic as the *gobaz shemaglee*, or noble elder. Desmond listened to the offer, and, annoyed at this patent blackmail, said very loudly: 'Ten cents is far too much. Offer him five!' Whereupon the *balabat*, who had obviously caught the gist of this counter-offer, turned on his heel and began to stalk off, surrounded by his glowering

Fig. 14.6 Molar of Pliocene *Elephas recki* used to build stone enclosure for baby goats, Afar Desert, Ethiopia, 1975

warriors. Fortunately at that moment Betty came rushing out of her tent and called out: 'Desmond, call him back! I'm worth ten cents even if you're not!' Face was restored, although by now the price per head had risen.

In the Afar Desert of Ethiopia, the fossils of long extinct Pliocene animals are so abundant and so well preserved that the local Afar tribesmen use them to build circular walled enclosures to protect their young animals from predatory nocturnal carnivores such as jackals and hyenas. In one such wall I saw a complete *Elephas recki* molar, identified by Dr Tim White as late Pliocene or roughly three million

Fig. 14.7 Volcanic crater ('Devil's Island') in the Red Sea near Djibouti. *Photo* Françoise Gasse, 1975

Fig. 14.8 Fossil skull of early Homo from Bodo, Afar Desert, showing cutmarks. *Photo* Tim White 1981

years old (Fig. 14.6). At that time the Afar was far less arid than it is today. Until about four million years ago, a vast freshwater lake occupied what is now the middle Awash valley, and early hominids foraged for plants and may have hunted occasional small animals in the woodlands adjoining the lake. Near the top of the thick alternating sequence of these lake sediments and lake margin swamp clays, a cindery volcanic tuff 40 cm thick, and dated by three independent methods to about four million years in age, shows that the lake had persisted for several hundred thousand years until that time, but vanished thereafter, for reasons that remain obscure. In a region like the Afar, where earthquakes are common and volcanic activity far from dormant (Fig. 14.7), it is often very hard to discern whether the causes of local and regional environmental changes are tectonic, volcanic or climatic in origin.

We did make some rare and unusual discoveries, both before and after the main field season. Some years earlier another team working near Bodo had recovered the skull of a male human transitional between *Homo erectus* and *Homo sapiens*, later shown to be about five hundred thousand years old. The skull was encrusted with calcium carbonate and was stored in the museum in Addis Ababa. Tim White obtained permission to clean the skull, which he did with great skill. To his surprise, he found a series of cut marks along the bones of the skull and inside the eye sockets (Fig. 14.8). He realised that the cut marks represented deliberate and careful defleshing, possibly for some ritual purpose, and presumably using sharp blades of obsidian or volcanic glass. He later examined skulls of about this age collected from other parts of the world and found that a number of these skulls displayed similar patterns of cut marks. This fortuitous discovery thus provides evidence of one of the oldest forms of ritual among early humans.

Another great discovery was finding the fossilised bones of part of a skull and of a small femur (thigh bone) at a depth of about 8 m beneath the 4 million years old volcanic tuff. These bones belonged to a hominid with a small brain and the ability to walk upright. The most likely candidate is *Australopithecus afarensis*, the species to which the well-known fossil 'Lucy' belongs.

As I explained earlier (see Chap. 9), we were not able to resume work in the Afar the following year, so Desmond and I returned to the Son valley in India in 1982. Tim White was the worst affected by the ban because he had resolved only to survey for fossil sites in 1981, marking their position with piles of stones. His plan was to return and excavate the following year. It was to be over a decade before he and Desmond were once more granted permission to return and excavate. I should add that Tim and Desmond had always been scrupulous in helping to train young Ethiopian archaeologists and palaeontologists (fossil specialists), and had helped fund the construction of the museum and research laboratories in Addis Ababa. Unfortunately, fossils attract some nasty people, and the ensuing fallout adversely affects everyone else. This type of work demands immense commitment, great patience, and diplomatic skills of a high order as well as a very high degree of professional competence and scientific honesty.

Chapter 15
Rajasthan, India (1983)

In December 1982 Desmond had assembled a formidable team of specialists in Addis Ababa to continue the work in the Afar that had proven so promising the previous year. A very wealthy American lady with a keen amateur interest in archaeology had very kindly offered to put up our party as her guests in the Hilton Hotel in Addis while we prepared for the fieldwork and Desmond had agreed to this generous invitation. It was a level of luxury to which we were quite unaccustomed and not without its snags. The telephones in every room were bugged, as we discovered very soon. One morning we were all given a couple of hours notice to pack and leave, as were all the other hotel guests. While Don Adamson and I were hastily cramming our sparse belongings into our backpacks, a couple of secret police arrived to remove the bug from our bedside telephone. The reason behind all this urgent activity eventually became clear. A large contingent of Zimbabwean government

M. Williams, *Nile Waters, Saharan Sands*, Springer Biographies,
DOI 10.1007/978-3-319-25445-6_15

officials was about to arrive in Addis and the Hilton had been commandeered for their sole use. Somewhat later that year the Ethiopian military dictator and president Haile Mariam Mengistu fled from Ethiopia with his retinue of bodyguards and family and took up residence in Harare, the capital of Zimbabwe, where he soon acquired a reputation among the local people for brutal treatment of his servants.

Our party settled into less up-market quarters while Desmond sought in vain for permission to proceed down to the Afar. I have told earlier how Don, Peter Jones and I made a quick decision to fly on to Khartoum and resume fieldwork along the White Nile (see Chap. 9). One person whom Desmond had invited to be part of our team was Professor Virendra N. Misra, a distinguished Indian archaeologist with a broad knowledge of all aspects of prehistory and an experienced site excavator. I liked Virendra—a gentle and scholarly man with a fine sense of humour. Out of the blue, he suggested that I could join him and his team for some fieldwork in the desert of Rajasthan when on my way back from Sudan to Australia. Just a matter of catching the Jodhpur Mail from Delhi railway station to Didwana (Fig. 15.1), arriving at dawn, and then asking a *tonga wallah* to take me to the site where they were working out in the desert. This I duly did, arriving one cold winter's morning as the sun was rising to see a small group of well-muffled individuals huddled out in the open over a tiny fire. I had reached my destination.

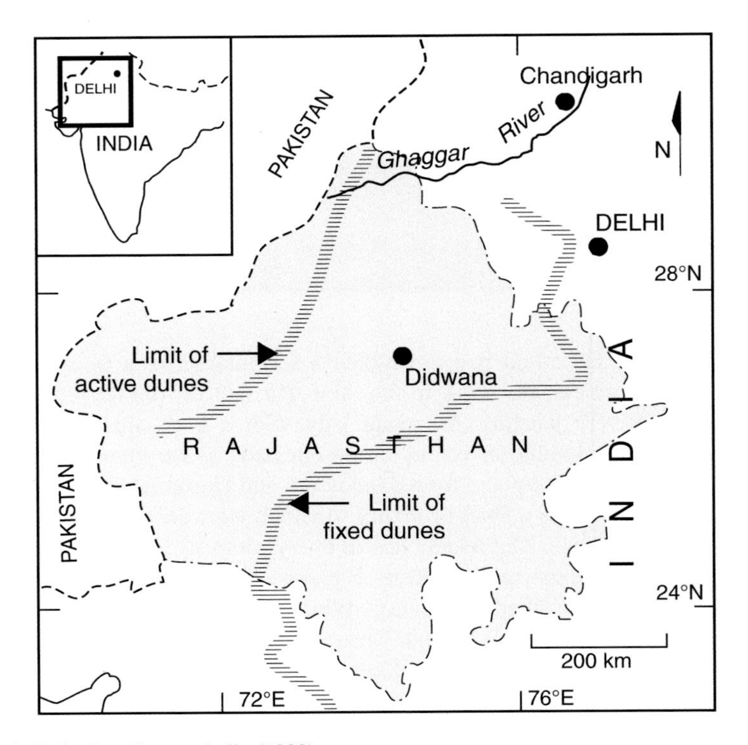

Fig. 15.1 Rajasthan Desert, India (1983)

For some years Professor Misra and his friend and colleague, Professor S.N. Rajaguru, who was India's most experienced Quaternary geologist, had been investigating sites relating to the Indus Valley Culture. This great urban and agricultural civilization arose along the Indus River and its major tributaries some five thousand years ago and is perhaps best known for the cities of Harappa and Mohenjo-Daro. One of the many puzzles related to this civilization is why it declined quite suddenly about 3500 years ago. Was it because of the invasion of Aryan warriors from the steppes of Persia and Afghanistan at about that time? Or was it because the climate had become much drier quite suddenly, as my Australian National University friend and colleague, Dr Gurdip Singh, had long argued, based on his studies of pollen grains in the sediments of some of the Rajasthan lakes? Or was there another reason altogether? It all rather reminded me of the last poem written by W.H. Auden (1907–1974), entitled *Archaeology*, in which he concluded that 'Knowledge may have its purposes, but guessing is always more fun than knowing.'

Mohenjo-Daro is located on the Indus River and Harappa is on the Ravi River, but some of the Indus Valley sites were located on a once active branch of the Ghaggar-Hakra River, which was in part fed from the former headwaters of the mighty Yamuna and other Himalayan rivers. The Yamuna no longer flows south but now flows east and joins the Ganga at Allahabad, nearly a thousand kilometres to the east. Virendra had come up with the idea that prehistoric earth movements had diverted the headwaters of the rivers that once fed the Ghaggar-Hakra River so that it dried out in its lower reaches. Certainly, it is an unremarkable stream today in its upper reaches, easy to wade across, and quite dry downstream. In Hindu tradition, the ancient Ghaggar-Hakra River was the mythical Sarasvati River, which is said to flow underground to the present junction of the Yamuna and Ganga rivers at Allahabad—today a place of pilgrimage for devout Hindus from all over India. In

Fig. 15.2 Village men and women who helped us excavate the deep trench near Didwana, Rajasthan Desert, India, 1983

order to test his ideas of river diversion and river desiccation, Virendra was in the process of mapping and dating sites related to the Indus Valley Culture and was finding that they became progressively younger to the east, out on the vast alluvial plains of the great Ganga River.

Meanwhile, there was other work to be done at a spot out in the desert near Didwana salt lake with the thoroughly unromantic name of 16R. Here Virendra, Rajaguru and graduate student Makhan Lal had enlisted the help of men and women from a nearby village to excavate a very deep trench through a dune (Fig. 15.2). Two of the younger village ladies did not think much of my digging efforts and proposed that I stick to measuring and taking notes. The trench ultimately went down to a remarkable depth of 18.3 m, thanks to a wooden tripod and pulley-system rigged up by the ever-resourceful Makhan Lal (Fig. 15.3). The only reason we could go so deep was that every metre or so in the stepped trench we came across very hard layers of calcium carbonate. The appropriate geological term for these layers is calcrete and they were certainly not easy to dig through. We stopped at a depth of 18.3 m because even our cold chisels and sledgehammers could not prevail on the bottom layer of calcrete. Still, we had an unusually long sequence of prehistoric stone tools ranging from Lower Palaeolithic (Early Stone Age) at the base to Mesolithic at the top. When eventually dated by my friend and colleague Professor Ashok Singhvi using optical dating methods, we realised that our trench spanned about two hundred thousand years and showed evidence of about a dozen climatic cycles ranging from dry (with deposition of wind-blown sand) to wet (with soil formation and in due course calcrete formation). The source of the calcium carbonate was most likely from wind-blown dust, so that my thoughts returned once more to the enigmatic mountain valleys in the Matmata Hills of Tunisia.

Fig. 15.3 Using the tripod and pulley erected over the base of the deep trench near Didwana, Rajasthan Desert, India, 1983

One cold January morning a group of about twenty-five itinerant snake catchers had set up camp on the flank of a dune about a kilometre north of the deep archaeological trench we were excavating through a sequence of alternating wind-blown sand and resistant bands of massive calcium carbonate that were invaluable in stabilising our step trench. I wandered across for a chat. The snake catchers showed me some dried specimens of the plant that they harvested from the dunes as an antidote to snake bite. They travel about 10 km a day with their asses, goats, sheep, dogs and chickens, covering about 1600 km each year through Rajasthan.

A few days later, with another member of our team, French archaeologist Claire Gaillard directing operations, we were excavating a nearby open site strewn with sporadic Lower Palaeolithic hand-axes and cleavers scattered through a bed of green alluvial clay. To me it was suggestive of episodic deposition of the stone tools by itinerant Lower Palaeolithic hominids during progressive sedimentation at this former flood plain site. These tools had saw-like cutting edges and were all-purpose tools; they could be used for skinning, cutting through tendons, or removing bark from trees to get at the inner and often edible layer of cambium. The desert dunes and

Fig. 15.4 Villager about to lop branches from a *Prosopis spicigera* tree, Rajasthan Desert, India, 1983

Fig. 15.5 Pile of branches obtained after lopping the *Prosopis spicigera* tree, Rajasthan Desert, India, 1983

surrounding plains of Rajasthan thus contain a record of intermittent human activity extending back from the present to at least two hundred thousands years ago.

The Rajasthan desert is home to the Bishnoi tribe who have a long and impressive history of conserving the natural world. They are vegetarian and do not allow hunting of the desert antelopes and other animals. They will never cut down a living tree but use dead wood or dried animal dung for fuel. On a smaller scale, the villagers living near the 16R dune site practised their own form of conservation. One day I came across an elderly man armed with a wooden fork and a long pole with a cutting blade attached at the end. He was lopping leafy branches from the *Prosopis spicigera* trees growing on the dunes and collecting it in piles as fodder for his goat (Figs. 15.4 and 15.5). Once the goats had eaten their fill the remaining branches would be used for fencing. When they became too dry and brittle to be useful as fences they served for fuel.

Water is essential to life in deserts and we were no exception. Our water came from a concrete cistern and the level was very low. Worse still, dead rats were floating in the water, which we boiled assiduously. Virendra was convinced that if we were to throw in a handful of potassium permanganate (Condy's Crystals) this would sterilise the water. Perhaps it did, but it turned everything in the water an unpleasant shade of pale purple. I was concerned about typhoid and said we needed to persuade whoever controlled the water supply to flush out the bad water and replenish it with fresh water. It transpired that a local dacoit (bandit) leader somehow had control of our water but he did oblige. Later, with our main tasks completed, we travelled through parts of this most attractive of deserts, checking out a number of good geological sections, naturally, but also enjoying the wonderful architecture and colourful people.

In their widely cited popular volume *Climates of Hunger* (1977) Reid Bryson and Thomas Murray claimed that the desert of Rajasthan was only a few thousand years old and had been caused entirely by human mismanagement by the people who built Harappa and Mohenjo-Daro some four thousand years ago. They argued that the dust produced by removal of the native vegetation through overgrazing and excessive cultivation weakened the monsoon and resulted in sustained drought and so to desertification on a grand scale: 'They helped make a dustbowl out of a bread basket, and have kept it that way.' Bryson was a well-known climatologist and Murray a professional scientific writer. Neither man had actually carried out any fieldwork in Rajasthan, but both seemed content to theorise on a grand scale. Our work at the deep 16R trench has shown that sand has been blowing around in this landscape for at least two hundred thousand years. That means that the Rajasthan desert is at least that age also. As I have written elsewhere: 'the best way to learn is to go out and discover for yourself through fieldwork. It is also a lot more fun. As always, an ounce of practice is worth more than a pound of theory…but a little theory can and does help!'

Chapter 16
Somalia (1988)

In 1988 Steve Brandt invited me to join his team for some fieldwork in Somalia. I had worked with Steve in Ethiopia and India and had also been his graduate advisor when he was a doctoral student at Berkeley. Steve was now in the department of anthropology at the University of Florida and had been excavating archaeological sites in Somalia for a number of years. The Somali government had invited him to help carry out an environmental impact assessment in the Jubba Valley (Fig. 16.1). The Jubba flows from the highlands of southern Ethiopia through a series of limestone gorges in western Somalia and is the only big river in Somalia to reach the sea. The Somali government planned to build a dam near a place called Bardera.

The stated aims of the dam project were to generate hydroelectric power for the capital Mogadishu and to provide a reliable supply of water for local irrigation. In

M. Williams, *Nile Waters, Saharan Sands*, Springer Biographies,
DOI 10.1007/978-3-319-25445-6_16

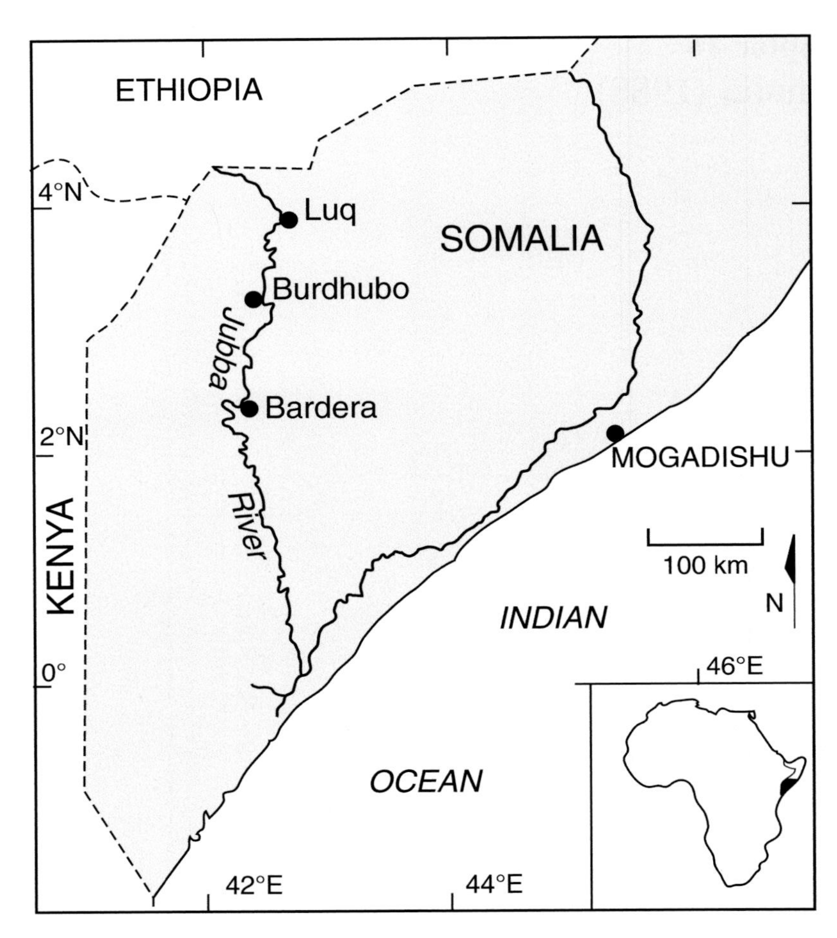

Fig. 16.1 Jubba Valley, Somalia (1988)

order to secure World Bank funds the Somali authorities needed to carry out an environmental impact assessment (EIA) of the area likely to be submerged by the reservoir and adjoining areas. This EIA process would involve a survey to see if any prehistoric or younger heritage sites would need to be either protected or excavated. My task was to provide some environmental context for the archaeologists by carrying out a month-long exploratory survey on foot of the landforms, soils and alluvial sediments along the eastern Jubba valley between the proposed dam site at Bardera and the small town of Luq over three hundred kilometres to the north.

There was a weekly flight from Nairobi to Mogadishu. In Nairobi I visited the National Museum and had a lengthy chat with Richard Leakey, who was the Museum Director and also knew Steve Brandt. Richard was awaiting a kidney transplant from his brother Philip at the time and was remarkably composed and very good company despite his worries.

I arrived in the capital city of Mogadishu to be briefed by Steve Brandt. At that time Mogadishu was an elegant coastal city with many ornate and attractive buildings dating back to Italian colonial times. Steve immediately advised me not to wander the streets of the capital wearing khaki shorts or shirt or I might be fatally mistaken for some of the mercenary pilots brought in by the government of President Siad Barre to bomb the northern city of Hargeisa, then attempting to break away from Somalia. It was not an auspicious start (Siad Barre was forced from office three years later in 1991 as a result of fighting between the big clans and the internecine warfare between various militias, which has endured to this day. At the time of my visit in October 1988 the seeds of unrest were spreading throughout the country and things were starting to fall apart).

Our small party set forth from the village of Bardera. I had two camels to carry our supplies and my portable rock-drill, two local camel men (Ibrahim Isaak and his assistant Mohamed Nur), an armed askari (Hussein Hassan), my Somali counterpart (geologist Abdurahman Mohamed Kamil) and our cook (Mohamed Abdulla Mohamed). As we trudged north my aim was to explore each of the main side valleys for any dateable sediments I could find. The long drought had broken in the headwaters and the river was in flood, its waters coloured red-brown because of all the silt and clay washed down from the Ethiopian highlands and carried in suspension.

As we progressed upstream we witnessed one of life's unforeseen tragedies. Late one afternoon I went down to the river to fill a bowl with water for washing. It had been a very hot day scrambling through some dense thorn scrub. I spotted a human body floating slowly past. We alerted some local Somalis who put out in a wooden dugout canoe and retrieved the body of a lad in his mid teens who had been in the water about two days. There was no sign of trauma on the body. I arranged with the local sheikh (headman) for the boy to have a proper Muslim burial and returned to the river in time to see a second body float past. We were unable to retrieve this body and by now it was dark. I took my cue from my two camel men, who seemed unperturbed, and we proceeded upstream, checking out the geology on the way. Two days later we reached a place called Burdhubo, where there was a bridge across the river and a refugee camp on the opposite side of the river. Here I received a great welcome and learned the full sad story. The drought in western Somalia had driven these desert nomads to settle next to the river, which was now in flood. The boy had gone to fetch water, slipped and fell into the river. His father jumped into try and save him. Both were swept away.

A day later we came upon a very irate Somali farmer and a young American archaeologist. The latter had put down a test trench in one of the fields belonging to the farmer without seeking permission. The situation was in danger of turning violent. I called on the local sheikh and after an exchange of elaborate greetings we both agreed that young men could be stupid and hot headed. To soothe ruffled feelings I bought a goat and we agreed to have a meat feast that evening. I insisted that our happiness would be incomplete without the presence of all concerned, including the sheikh from whom I had bought the goat. He was reluctant (I was soon to find out why) but finally agreed. It proved to be the toughest old goat I have ever eaten. My jaws ached for days afterwards!

A few days on we made a long forced march across some jagged limestone country. On reaching the bitter springs we found large numbers of camels being brought into drink. I replenished my water bottle, unaware that the spring waters were really a dilute form of Epsom salts, and an effective aperient. Nemesis struck later that day. As I was lying on some angular blocks of limestone gazing up through dead gnarled branches at the relentless sun, I could not help asking myself, like Bruce Chatwin when he was once lined up before a collection of armed and drunken soldiers during a coup in West Africa, 'what am I doing here?'.

Providentially, one of the senior American archaeologists, Nanny Garder, turned up out of the blue, and gave me water with electrolytes to drink, which did the trick. Later that day Kamil and I set forth for a nearby cave to try and obtain a core sample from the dripstone curtain inside the cave. I was operating the rock drill and Kamil was pumping water from a jerry can to cool the drill bit. He suddenly screamed, clutched his head and dropped the jerry can. A moment later I realised why as I too was attacked by a swarm of savage African bees. We beat a hasty retreat, extracted the stings, and decided to return at dawn when the bees would be cold and inactive. Fortune favoured us and we obtained a useful core. Later analysis would tell me when the climate in this area had last been wetter, allowing me to build up a picture of past environmental changes.

By now I was becoming increasingly puzzled by the absence of high-level alluvial terraces running parallel to the present river or extending in along the side valleys. The Jubba River originates in the highlands of Ethiopia, and I was very familiar with the history of lake level fluctuations spanning the last forty thousand years and more in that region. The best I could find was in a side valley called Tog Bali, where there was a series of low silt terraces up to a few metres high. Close inspection showed that they contained the sub-fossil shells of aquatic snails, including *Bulinus truncatus* and *Biomphalaria pfeifferi*, which I knew all too well from my days in Sudan. Both of these snails are vectors of the *Schistosomiasis* or *Bilharzia parasite*. In addition, the terraces contained a suite of Later Stone Age and younger stone artefacts. I concluded that these pathetically small terraces might be the counterparts of the spectacular flights of late Quaternary lake shorelines in the Ethiopian Rift, but I remained puzzled. The answer came during October 20 and 21, when intense downpours added to an already swollen river and caused all the small side valleys to become flooded. The red silts and clays brought down in suspension from Ethiopia were laid down as slack-water deposits in the shallow limestone valleys aligned roughly perpendicular to the main valley. The limestone tributary valleys were so highly fissured and permeable that major floods were not able to attain elevations of more than a few metres above the level of the flood plain adjoining the main river, thereby accounting for the absence of high-level alluvial terraces.

If this conclusion was indeed correct, it meant three things. First, the proposed reservoir would never attain the height estimated by an earlier team of East European engineers, who seem never to have checked out the local geology.

Fig. 16.2 Stick framework of portable hut, Jubba Valley, Somalia, 1988

Fig. 16.3 Completed hut, Jubba Valley, Somalia, 1988

Second, the valley would be flooded to a shallow depth, but one sufficient to submerge the little arable land cultivated by the local people today (Figs. 16.2, 16.3, 16.4 and 16.5). Third, there was also a strong possibility, nay, probability, that the parasitic disease *Schistosomiasis* (*Bilharzia*) would spread throughout the valley. The parasites do not kill you but you lose blood fore and aft and can become very weak and anaemic.

I had learned that Tom Gresham, who with Steve Brandt was co-director of the archaeological surveys, was working along the opposite bank not far from where we had reached. I resolved to cross the river and discuss my concerns with Tom and

Fig. 16.4 Somali man, women and camel, Jubba Valley, Somalia, 1988

Fig. 16.5 Somali woman and her children, Jubba Valley, Somalia, 1988

Fig. 16.6 Dugout canoe, Jubba River, Somalia

his cohort of Somali engineers. The local dugout canoes are designed for slim and slender Somalis, so there was not much freeboard (Fig. 16.6). The current took us about a kilometre downstream. I sat down with the Somali engineers and outlined my concerns. They looked grave and said I should get to Mogadishu as soon as possible and discuss it with their government minister in charge of the Bardera dam project. This I did and was asked to put my concerns into a confidential appendix to my main report.

Kamil had been invalided back to Mogadishu—the rough walking had damaged his feet. His replacement was a portly engineer, Abdulkadir Ibrahim Derow. His

Fig. 16.7 Sunset after rain in the Jubba Valley, Somalia, 1988

feet suffered too but he proved stoical. My two camel men complained to me about these urban softies and could not understand their role—which was ostensibly to help me but in reality to keep an eye on me, and to report back to their bosses in the capital.

For the second leg of the long walk we started from Luq in the north and moved south. Progress was slow at the start because it took us several days to locate our camels. Once we finally started we realised that we would have to move swiftly. A blasting sandstorm followed by a sudden downpour convinced us that contrary to all expectations the rains were on their way and would come early (Fig. 16.7). Camels do not fare well on wet slippery clay surfaces and can easily break their legs. We decided that from then on we would start early, no later than 4 am, and walk rapidly southwards, limiting ourselves to one quick meal at night. I made what observations I could but our prudence paid off and we reached Burdhubo almost out of food but with our camels intact. Another gusty downpour blew all our small tents down except for one in which we kept the remains of our food. Drenched to the skin, we were hanging onto the main tent pole to stop the wind from blowing the food tent away. I became aware of something delicately plucking at the hairs on my legs. Looking down I saw a small wet monkey. The monkey adopted me and slept on my sodden sleeping bag that night. At dawn it shinned up a tree, peered across the river, and scurried across the bridge. By now we were out of sugar for our tea, and were feeding a few others from our wider group, who were also awaiting transport back to Mogadishu. I offered to walk to the nearest camp, a few kilometres away, to buy some sugar. By now I was in walking mode and the distance did not seem too great after a total walk of nearly 700 km. On arrival at the small camp I purchased the sugar and was enjoying a glass of sweet tea when my small monkey appeared on a fence. I stood up and made suitable monkey noises. It sailed through

Fig. 16.8 Journey's end: the author, Ibrahim Isaak, Hussein Hassan, Mohamed Abdulla Mohamed and Mohamed Nur, Somalia, 1988. *Photo* Abdulkadir Ibrahim Derow

the air, landed on my shoulder, slipped down and continued grooming my legs. The local Somalis initially tried to tell me the monkey was dangerous and would bite. They were amazed. One elderly man with a lovely white beard stood up, cleared his throat, and addressed the crowd: 'This good man is the father of this monkey; and this monkey is the child of this good man.' Having delivered himself of these compliments he sat down and resumed chewing his mildly narcotic *khat* (pronounced like 'chat') leaves.

Our vehicle eventually arrived. I said farewell to Hussein, our gentle *askari*, and to our two tough camel men (Fig. 16.8). There was one flight a week from Mogadishu to Nairobi. At the airport the customs men tried to confiscate my precious geological samples (they wanted money). I later learned that if they did not secure enough money for their bosses from the foreigners leaving Mogadishu they were sent to fight on the northern front and never returned.

There was a very sad sequel to my time in Somalia. About a year later a group of armed men from a rival clan attacked Bardera and I learned from Kamil that my former camel men (Ibrahim Isaak and his assistant Mohamed Nur) had died in the fighting. Since that time the country has split into three separate countries and the local people continue to endure the hardship caused by recurrent drought made worse by widespread lawlessness and continuing warfare.

Chapter 17
Inner Mongolia, China (1999)

After moving from the School of Earth Sciences at Macquarie University in Sydney to the excellent Department of Geography at Monash University in Melbourne (Fig. 17.1), I became more and more interested in questions of soil and water conservation. There were several reasons for this renewed interest in how to prevent or minimise land degradation. I had long been concerned about soil erosion and had monitored hill slope erosion in tropical Northern Territory and temperate New South Wales during my doctoral research. When I moved to Victoria I came across an even more insidious form of land degradation than soil erosion, namely, salt accumulation in the soil.

In order to appreciate the causes of this salt menace we need a brief bit of background explanation. Two centuries ago much of southeast Australia was covered in forest or woodland. With the arrival of settlers, soldiers and convicts

M. Williams, *Nile Waters, Saharan Sands*, Springer Biographies,
DOI 10.1007/978-3-319-25445-6_17

Fig. 17.1 Explaining to a group of Monash students some of the finer points of an Early Stone Age hand-axe made by *Homo erectus* in the central Sahara half a million years ago

from England, Scotland, Wales and Ireland, a massive wave of tree clearing moved across the landscape. Within less than two hundred years over half of the woodlands and forests that once covered the continent had gone. One consequence of these activities only became evident a century or more later. The eucalypts act as natural groundwater pumps, pumping water from around their roots and discharging it to the air as they transpire. The trees therefore keep the water table well below the surface of the ground. Remove the eucalypts and replace them with cereal crops like wheat and barley, and you will initiate a slow but steady upward rise in groundwater level. As the groundwater moves gradually towards the surface it dissolves any salt in the sediments through which it is rising. Once it approaches the rooting zone of crops, especially those most vulnerable to salt, there is a rapid decline in crop quality and yield, culminating in the death of plants in the worst affected areas.

Victoria was one of the prime agricultural regions in Australia most severely affected by the progressive increase in the salt content of soils and streams. On arriving there I threw myself with gusto into the work of the *Soil and Water Conservation Association of Victoria*. We set up a working group of highly experienced soil scientists, agricultural economists and agronomists (including Dr Robert van de Graaff, Frank Gibbons and Dr Bob Dumsday) to report on some of the more widespread forms of soil erosion within the State. By now I was becoming more and more involved in investigating the causes and consequences of dry land degradation (or desertification) in all its manifestations, as well as with finding possible solutions. Victoria led Australia with the Decade of Landcare initiative thanks to the foresight of Minister Joan Kirner, who later became Premier of Victoria.

Almost inevitably, my attention was turning increasingly to the question of drought. What were the primary causes of drought? Could changes in land-use cause a reduction in rainfall? Could we predict the likely magnitude of droughts in Australia (and elsewhere) by monitoring changes in sea surface temperatures? Why were major droughts (and floods) over the past several hundred years synchronous in regions as far apart as eastern Australia, peninsular India, eastern China, the Nile basin and northeast Brazil?

I took part in a number of workshops and conferences on drought and desertification held in Australia and at the United Nations Environment Programme (UNEP) headquarters in Nairobi, Kenya (Fig. 17.2). While in Nairobi at a UNEP conference I was asked to prepare a major report for UNEP and the World Meteorological Organisation (WMO) on the interactions between climate and desertification to be ready in time for the drafting of the *UN Convention to Combat Desertification*. I co-opted Dr Bob Balling, a lively climatologist from the State University of Arizona in Tempe, to help with this task, which we managed in enough time to have it rigorously reviewed during a week in Geneva in late 1993. It was subsequently expanded, reviewed very thoroughly by my friend and colleague Professor Peter Lamb from the University of Oklahoma and published by Arnold, London with WMO and UNEP as a book (Interactions of Desertification and Climate, 1996).

All of this somewhat theoretical work turned out to be a prelude to something more practical. In 1999 I was invited to go to the arid Alashan region of Inner

Fig. 17.2 At one of many workshops and conferences on desertification run by the United Nations Environment Programme in Nairobi, Kenya

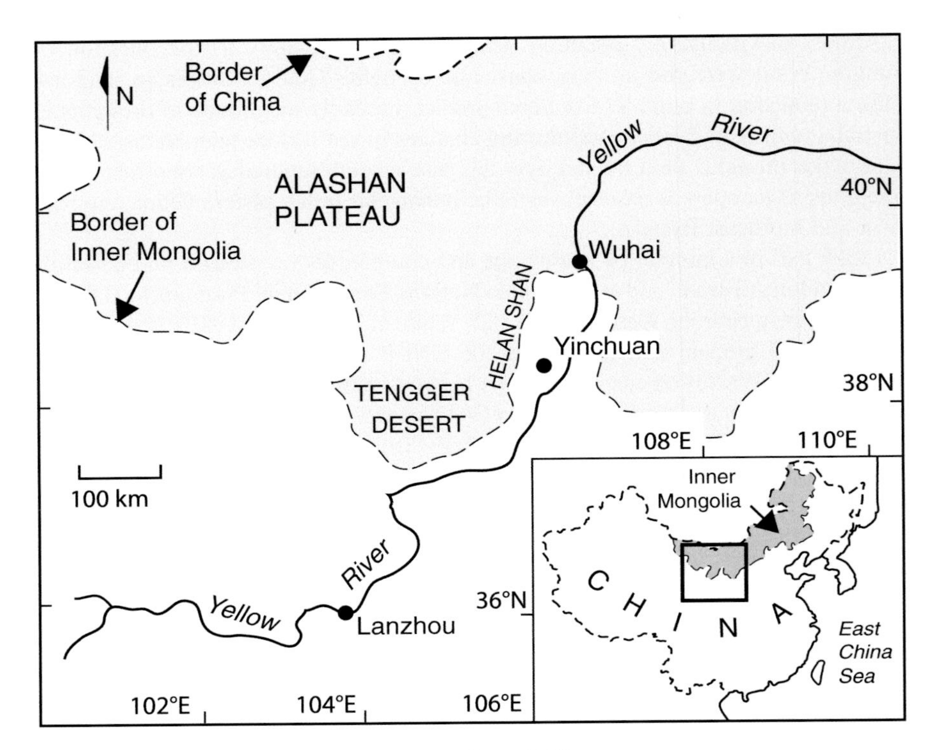

Fig. 17.3 Alashan, Inner Mongolia, northern China (1999)

Fig. 17.4 Distant view of mud-brick fort abandoned as a result of climatic desiccation some two thousand years ago in Xinjiang Province, northwest China, 1999

Fig. 17.5 Mud-brick fort abandoned as a result of climatic desiccation some two thousand years ago in Xinjiang Province, northwest China, 1999

Fig. 17.6 Mountain front, Taklimakan Desert, northwest China

Mongolia in northern China (Fig. 17.3) to report on desertification there as part of a joint initiative between the governments of China and Australia. During previous working visits to China I had spent some time in the Taklimakan Desert of Xinjiang Province in the far northwest (Figs. 17.4, 17.5, 17.6 and 17.7); drilling in Inner Mongolia's Lake Daihai for sediment cores; inspecting gully erosion on the Ordos Plateau in the northeast; and assessing soil conservation measures on the Loess Plateau of north-central China. I was therefore familiar with some of the problems facing these dry regions and so was happy to accept the invitation.

Fig. 17.7 Dry valley issuing from mountain range, Taklimakan Desert, northwest China

In Beijing I met up with Dr Wang Tao, Director of the Chinese Academy of Science Desert Research Institute in Lanzhou, with whom I had worked in Geneva in 1993, and with Professor Yang Xiaoping in the Institute of Geology and Geophysics, Chinese Academy of Science, Beijing. Xiaoping was a close friend of my PhD student Zhou Liping (Fig. 17.8), having been to school with him in China, and was also a younger colleague of the great Professor Liu Tungsheng, whose pioneering work in the Loess Plateau had long attracted international attention. So, among friends and like-minded scientists, the formalities took a couple of hours and not several weeks, as I had been warned to expect, and I proceeded swiftly to my field area.

Rising abruptly along the eastern boundary of my survey area was the Helan Shan range. Mantled in snow during the winter months, these mountains were clad in beautiful shady pine forests through which bubbled limpid streams. Mushroom collectors came into the forest every second day to fill their tall osier panniers with

Fig. 17.8 Chinese scroll prepared for the author on the initiative of my Chinese PhD student Zhou Liping: *Meng Chi Chang Cai* ('Great aspirations endure forever')

the many varieties of mushrooms that grew beneath the forest canopy and were pleased to show me their collections. Vast alluvial fans spread out from the foot of the mountain range and merged imperceptibly into extensive gravel covered plains. In the west and south there were great tracts of vegetated sand dunes and in the far north the presence of sporadic aquatic snail shell fossils bore witness to a geologically recent time of wetter climate when shallow freshwater lakes covered large areas of the land. In the very far west was the Badain Jaran Desert, which has the highest sand dunes of anywhere on earth and higher than any dunes on Mars. These remarkable dunes attain heights of nearly 500 m, or over twice the height of the active Saharan dunes I knew from my time in Libya, Niger, Algeria and Sudan.

Desertification was a rapidly growing problem in Alashan. An influx of economic refugees from some of the poorer and overcrowded provinces south of the Yellow River had led to a doubling of the human population and a tripling of the animal population in Alashan since the 1950s. As a result, trees had been felled for house construction and gullies had developed at the foot of the Helan Shan ranges. Formerly well-vegetated and stable dunes had been stripped of their protective plant cover by overgrazing and were now migrating eastwards and being blown into the Yellow River. Great swathes of the beautiful *saxaul* (*Haloxylon ammodendron*) groves between the dunes, which used to provide fodder and shelter for the great herds of two-humped Bactrian camels, were now dead or dying. Large parts of Alashan had become, or were becoming, a wasteland. The challenge was to identify the processes responsible for this land degradation and to put in train a series of measures acceptable to and fully understood by the local people to halt these processes of desertification and restore the land to its former productive condition.

Deserts are lands of extremes. Icy cold in winter, torrid in summer, they can also experience flash flooding on a dramatic scale. In early August 1999 we were driving our two 4 × 4 vehicles across the wide boulder-strewn bed of an ephemeral stream channel in the arid Alashan Plateau of western Inner Mongolia. The entire region had endured eight years of savage droughts and the local Mongolian herdsmen were becoming increasingly worried for their camels, sheep and goats. This ephemeral stream arose in a nearby granite mountain and had barely flowed for several years. The Australian team member (me) was joking with his Mongolian and Han Chinese counterparts that there was no need to fear drowning in a flash flood, when there was a sudden flash of lightning followed almost immediately by a peal of thunder, and the heavens unleashed a torrential downpour. Well over 300 mm of rain fell in the next 30 h, sufficient to replenish water levels in the piedmont wells and to sustain plant growth for at least three more years according to the elderly Mongolian farmer and his wife, in whose home at the foot of the mountain our drenched and sodden party had taken refuge.

At about the same time, over half a world away, unusually heavy rains fell onto the parched clay plains of the central Sudan, flooding the hollows between the dunes immediately east of the White Nile, triggering sheet floods from the low upland separating the valleys of the lower Blue and White Nile rivers, and causing breaching and overflow from the main canal feeding into the most enduring and successful large irrigation scheme in Africa: the Gezira Irrigation Scheme (see

Chap. 5). This was no coincidence and I was at last starting to make some sense of global patterns relating to extreme climatic events such as floods and droughts.

One consequence of the torrential rain was to convert the fine sands next to the small stream channels in the narrow valleys to quicksands. We learned this the hard way next morning when we set off to explore one of the side valleys. Our Mongolian guide, not from this area, walked boldly ahead, screamed, and began to sink. He was a big man and it took some effort to extricate him from the quicksand. His tiny wife, meanwhile, berated him roundly for being such a silly man.

Chastened, we took to the high ground and wound our way slowly over or around granite boulders. Another scream and our guide was now to be seen writhing on the ground clutching his leg. A quick inspection showed no sign of snakebite and no fracture. To have carried him back to the farmhouse would be hard—he was certainly no lightweight! I glanced at his wife, who obviously thought he was malingering, which was also my impression. Psychology was needed. In stern silence, I indicated to three of our party to hold him firmly by the shoulders and thighs. I opened out the saw on my Swiss Army knife and mimed that I was going to amputate above the knee. His wife immediately recognised and approved my bluff. Our guide gave me a look of horror, shook off those restraining him, and rose to his feet, his powers of locomotion miraculously restored. I felt a bit sorry for our lumbering guide, who was not having a very good day, but I was immensely relieved that we did not have to carry him all the way back to the farm over such rough terrain.

On returning to the farmhouse where we had spent the night we found a sizeable group of very happy Mongolian herdsmen talking with great animation to our elderly hosts. I asked what was being discussed with such vigour and was told that they were planning to erect a statue of Buddha in thanksgiving for the return of the rains. The farmer's wife turned to me with a sly and mischievous grin and said 'and the face will look a bit like yours!'

Chapter 18
Flinders Ranges, South Australia: Solving the Puzzle (1993–2007)

At the start of 1993 I moved from being Professor of Geography and Director of the Graduate School of Environmental Science at Monash University in Melbourne to be Director and Foundation Professor of the lively Mawson Graduate Centre for Environmental Studies at the University of Adelaide (Fig. 18.1).

The move also allowed me to launch a course on environmental management for government natural resource managers from Asia, the Pacific, Africa and South America. We ran this course jointly with the United Nations Environment Programme. The concept arose out of earlier discussions I had conducted in Nairobi with Dr Michael Atchia who was in charge of environmental education at UNEP. A popular component of this course was an excursion across the South Australian desert, proceeding from the spectacular Flinders Ranges (Fig. 18.2) to the mound-springs along the southern portion of the Great Artesian Basin fed by

M. Williams, *Nile Waters, Saharan Sands*, Springer Biographies,
DOI 10.1007/978-3-319-25445-6_18

Fig. 18.1 Among the dolphins in the Port River estuary, with graduate student Steve Vines, Adelaide, ca. 1996

water upwelling under pressure, to the vast gold, silver, copper and uranium mine at Roxby Downs and the opal mines at Andamooka, operated by individual miners. It was during these field excursions that my interest in the climatic history of the South Australian arid zone was aroused and I was particularly intrigued by what seemed to be widespread evidence of previously wetter conditions in the now arid Flinders Ranges (Figs. 18.3 and 18.4), which I resolved to investigate in detail.

The rugged Flinders Ranges are oriented in a north-south direction and are flanked to west and east by two huge salt lakes. Rainfall decreases from south to north and the northern ranges are quite arid. The ancient rocks that make up the ranges vary greatly in their resistance to erosion. Where they consist of mudstones and siltstones there are wide gently undulating valleys carved out of these soft rocks. Steeply rising ridges of hard quartzites, sandstones and limestones, which contain very ancient fossils dating back to about 550 million years (Fig. 18.5), rise abruptly from the edges of the valleys. The landscape is spectacular, particularly when the sun's rays cause an explosion of colour in the sandstone rocks, which glow a golden red and yellow. The Flinders Ranges were one of the favourite spots

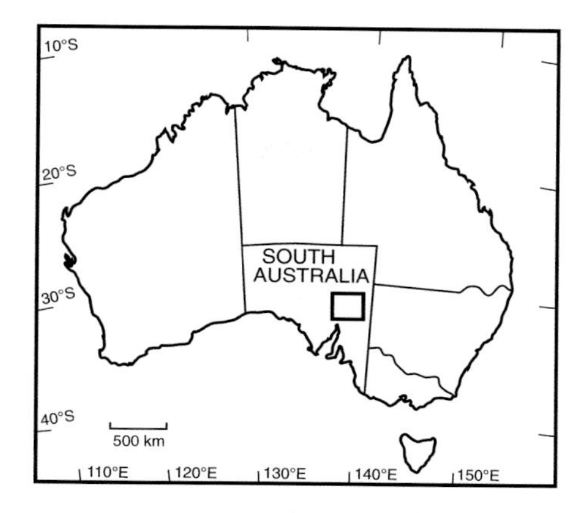

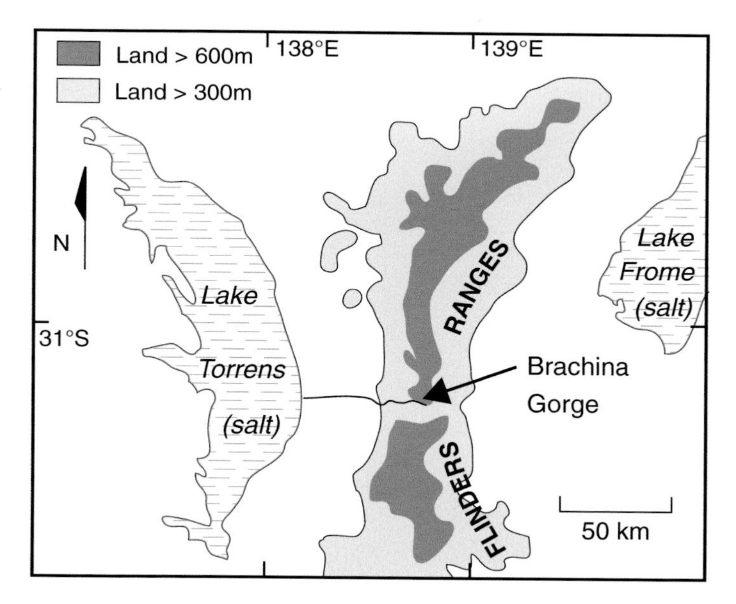

Fig. 18.2 Flinders Ranges, South Australia (1993–2007)

of the great landscape artist Sir Hans Heysen, who immortalised the gum trees growing in this majestic land.

A narrow gorge called Brachina Gorge runs at right angles from east to west through the central ranges. It is one of many such gorges. The walls of this gorge consist of spectacular and resistant ridges of quartzite, sandstone and limestone. In places the gorge narrows to a few score metres in width. Near the centre of the range the gorge suddenly widens out into a broad gently undulating valley underlain by easily eroded siltstone mantled in a layer of soft grey silty clay up to twelve metres thick. One remarkable exposure of this clay is located at the point

Fig. 18.3 Flinders Ranges, South Australia

Fig. 18.4 Flinders Ranges seen from the West, South Australia

where the gorge widens. The lower few metres are massive and contain unbroken shells of freshwater snails and tiny bivalve crustaceans called ostracods. Immediately above them the clays are finely laminated and indicate deposition in either a lake, a pond or under a very low energy regime.

My curiosity was aroused the moment I saw this section. Many questions came to mind. First, how old were the clays? Our subsequent work, buttressed by a battery of radiocarbon and optical dating ages revealed that they had been deposited between about thirty and fifteen thousand years ago, by gently flowing water in what seemed to be a wetland. The only problem was that this was a time of extreme

Fig. 18.5 Author inspecting a 550 million years old stromatolite fossil, Flinders Ranges, South Australia, 2008. *Photo* Frances Williams

regional aridity. In addition, the present ephemeral stream channel within the gorge seldom flows, but when it does, during rare summer downpours, it carries a traction load of gravel. No clay is accumulating today. Another puzzling feature is that despite rigorous searching, we have never found a single vertebrate fossil or indeed a single prehistoric stone artefact within these now very well dated deposits.

I began the search for an explanation for these enigmas in the trusty company of two old friends: Don Adamson, with whom I had worked in Sudan and Ethiopia, and John Chappell, who had been a PhD student with me at the Australian National University in the 1960s and had remained a close friend ever since, and two energetic and versatile PhD students: Peter Glasby, a science teacher in a Steiner School near Adelaide, and David Haberlah from Germany. Peter later worked with me in Kenya and in the Himalayan foothills in India, and David worked with me in the Sudan.

Like many geologists before us, we were initially misled by the very fine horizontal laminations and by the presence of the shells of aquatic snails and ostracods into believing that we were looking at sediments laid down in a former lake. We therefore searched the gorge for possible natural dams in the form of spring deposits of tufa that may have formed a barrage along narrow reaches in the gorge and found some possible contenders. However, the grey clays continued well beyond these putative tufa dams. We had the nagging feeling that we were missing something and so decided to carry out a detailed theodolite survey of the grey clays within the entire length of the gorge, tied into very precisely surveyed bench-marks giving location and elevation. Somewhat to our surprise we found that the surface of the grey clays sloped gently to the west, parallel to the present-day rocky channel floor, with a gradient of about 1 in 87, which is incompatible with deposition in a lake. One possibility we considered and quickly rejected was that there had been a lake but that later earth movements had caused tilting of the bed of the former lake to the west. To achieve this would have required unbelievably high rates of tilting, and would require that the older sediments were more steeply tilted than the younger ones, which was not the case.

We very soon concluded that we were dealing with a former wetland. But where had the clay sediments come from and what had allowed the wetland to persist during a time of regional aridity, when dunes were active and lakes were drying out across the continent? Spurred on by this apparent contradiction, we continued to assemble the missing pieces of the jigsaw.

During hikes along the ridge crests I had noticed that a thin layer of red-brown silt occupied cracks in the rocks regardless of the type of rock beneath. The silts must originally have been laid down by the wind as a mantle across the landscape. I was immediately minded of the desert dust deposits we had investigated in the Matmata Hills of Tunisia (see Chap. 12). Could we be dealing with a similar set of processes here in the Flinders Ranges?

Geochemical and particle size analyses strongly supported our notion that the fine-grained deposits within the valley bottoms were at least in part derived from former wind-blown dust mantles that had been washed off the hill slopes during times when the winter cloud base was lower and the rains were more gentle than today.

We also knew from other work that temperatures at this time (thirty to fifteen thousand years ago) were much colder than they are today, so that evaporation would have been much reduced, allowing a wetland to persist. We were aware from work carried out on ice cores from Antarctica and Greenland that the concentration of atmospheric carbon dioxide was very much lower at this time, which would have favoured grasses rather than trees. If the mighty River Red Gums (*Eucalyptus camaldulensis*) were knocked out, and so the natural ground water pumps were no

Fig. 18.6 Reworked wind-blown dust in the central Flinders Ranges, South Australia

Fig. 18.7 Dr Yoav Avni, Frances Williams and Bill McIntosh of Gum Creek Station near Blinman discussing causes of recent erosion, Flinders Ranges, South Australia, 2008

longer there to keep the water table from rising, groundwater would reach the surface and provide further moisture to sustain a permanent wetland.

Climate change brought about the demise of the wetland (Fig. 18.6). With the return of the summer monsoon about fifteen thousand years ago there was a dramatic change in the rainfall regime. The gentle winter rains now gave way to intense summer downpours, which led to incision into the alluvial silts and clays and the onset of widespread gully erosion within the ranges. As temperatures warmed up again, evaporation increased, so that any wetland remnants would have dried out for good. Furthermore, there was a rapid increase in the atmospheric concentration of carbon dioxide, so that the tables were now turned and trees resumed their dominance over grasses. The advent of the River Red Gums along the valley bottoms restored the natural groundwater pumps, so that the level of the water tables fell and the supply of groundwater was curtailed. The abrupt return of the summer monsoon about fifteen thousand years ago was also evident in the Nile valley (see Chap. 9) as well as in India and China, so that events in the Flinders Ranges were governed by a major reorganisation of the global atmospheric circulation system. It is also worth noting that the inception of gully erosion within the ranges had nothing to do with humans and everything to do with global climatic fluctuations. We humans may well have accelerated the processes of gully erosion since then (Fig. 18.7), but we certainly did not start them.

We had solved the Flinders Ranges wetland enigma, at least to our satisfaction, but that still left other important questions unanswered. Why were there no signs of human occupation associated with the wetlands at this time? Why had we been unable to find any fossil remains of wallabies, kangaroos and other animals within the wetland sediments? The search for possible answers continues…

Fig. 18.8 Skull Rock fossil waterfall, Flinders Ranges, South Australia

Humans have certainly been present within the ranges in prehistoric times as is evident from the scatter of stone flakes and grindstones sitting within the sediments that overlie and so post-date the wetland silts. Along the western margins there are human burial sites well over two hundred years in age, one of which was partly exposed in a stream bank section I discovered late one afternoon with Bryan Cock, one of my former graduate students. We alerted the traditional owners and showed them the site. After a very moving smoking ceremony during which one of the elders wafted the smoke from burning eucalypt twigs over us to protect us from any harm the eroding bank was protected with a palisade of large river cobbles.

A fossilised waterfall made up entirely of tufa and known locally as Skull Rock (Fig. 18.8) is another very curious feature of the landscape. The tufa is honeycombed with caves and hollows giving the appearance of a huge greyish-white skull when seen from several hundred metres away. Like the wetland, the waterfall was last active about twenty thousand years ago, when springs above the falls were providing enough runoff to allow water to flow down across the steeply sloping face of the waterfall, depositing thin layers of limestone. Until they reached the surface, the springs contained a considerable amount of dissolved limestone; on reaching the open air and losing some of their dissolved carbon dioxide, the spring waters reached saturation with respect to dissolved limestone so that the excess dissolved limestone came out of solution to form the tufa drapes.

One feature I was at a loss to explain was a limestone cave complete with stalagmites and stalactites located close to the summit of a very high hill. The cave must have formed when low in the landscape and in close contact with the water table. Current rates of erosion are very slow in this now arid environment, so that initial formation of the cave must have occurred millions of years ago; and yet the stalagmites were quite young, only a few tens of thousands of years old, so there is another mystery still to be solved.

Chapter 19
Kenya (1999–2003)

While I was working for Hunting Technical Services (HTS) mapping soils along the lower White Nile valley in Sudan, I spent my 1962 Christmas leave with some HTS friends on a visit to Kenya shortly before it achieved independence. We hired a car and visited the game park close to Nairobi. I was much taken with the country where the deep green colours of the vegetation and the bright red hues of the volcanic soils offered such a contrast to the harsh yellow grasses and grey clay soils of central Sudan. I climbed Mount Kenya in 1970 on my return from the Sahara to Australia and again in 1975 (Figs. 19.1, 19.2, 19.3 and 19.4) after fieldwork with Desmond Clark in Ethiopia. Once again, I found the landscape and people as attractive as ever.

M. Williams, *Nile Waters, Saharan Sands*, Springer Biographies,
DOI 10.1007/978-3-319-25445-6_19

Fig. 19.1 Summit of Mount Kenya from the distance, 1975. *Photo* Frances Williams

Fig. 19.2 Mount Kenya, Point Lenana, 1975. *Photo* Frances Williams

And so, when archaeologist Stan Ambrose invited me to join him for some fieldwork in Kenya in July 1999, I was happy to accept. I had met Stan during a visit to the University of Illinois at Champaign-Urbana in 1989 and we hit it off immediately. Stan had been a graduate student of my old friends Desmond Clark and Glynn Isaac at Berkeley (see Chaps. 7 and 8) and like me was interested in the possible climatic impact of the huge Toba volcanic eruption of 74,000 years ago. During 2005 we worked together on this problem in India (see Chap. 13). Stan's immediate aims for the Kenya fieldwork were to examine in detail the transition from Middle to Late Stone Age at a number of sites in the west Kenya Rift in the far

Fig. 19.3 Tarn at foot of Lewis Glacier, Mount Kenya, 1975. *Photo* Frances Williams

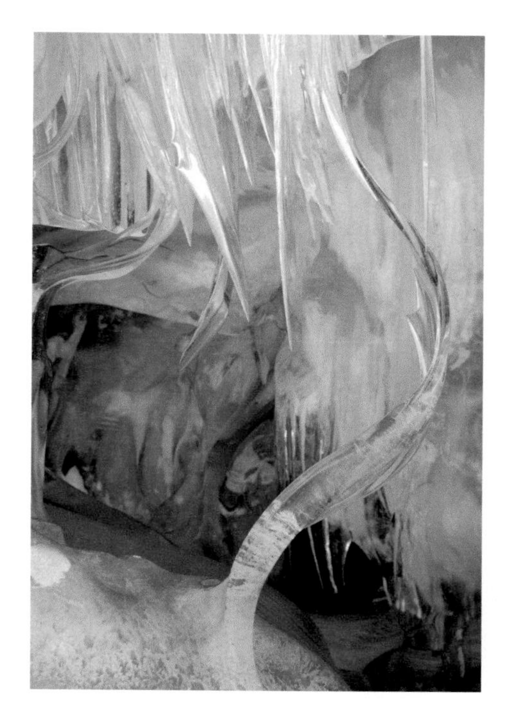

Fig. 19.4 Ice cave in Lewis Glacier, Mount Kenya, 1975. *Photo* Frances Williams

southwest of the country (Fig. 19.5). (In northern Spain and southern France the Late Stone Age or Upper Palaeolithic is renowned for its magnificent cave art, but the East African inception of the Late Stone Age promised to be much earlier). Stan had also located a very much older site at a place called Lemudong'o where the

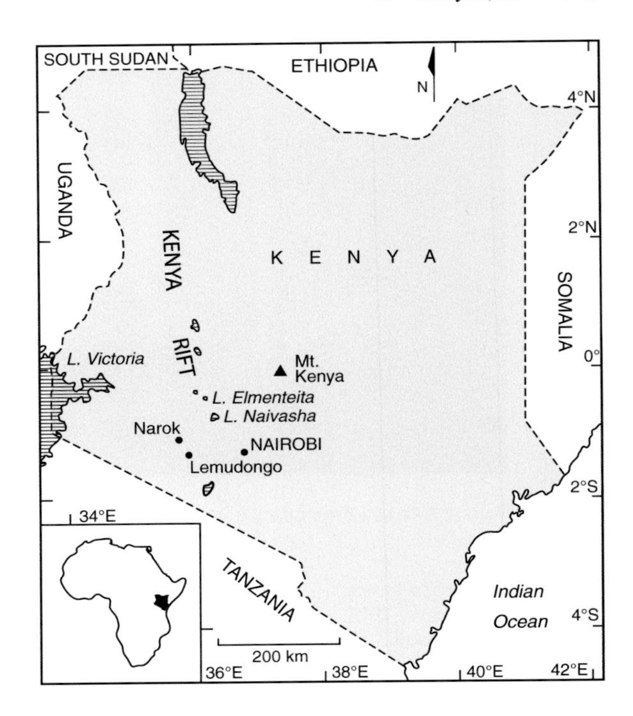

Fig. 19.5 Kenya (1999–2003)

abundance of fossil bones suggested a possibility of finding hominid fossils. Stan and I paid a quick visit to this site, scaled the cliffs above the fossil locality and collected samples of volcanic tuff. Dr Alan Deino at Berkeley subsequently obtained a potassium-argon age of six million years for these samples so we knew that the site would repay excavation during a later visit. Dr Leslea Hjulsko conducted these excavations a few years later and also ensured that the results were published in full and exemplary fashion, but she never found any hominid fossils.

My work in 1999 was cut short after I was hit in the left eye by a rock the size and shape of a golf ball. I was standing on a narrow ledge etched into a cliff 12 m above the ground watching one of our Kenyan graduates using a pickaxe to excavate through the debris above an archaeological layer we believed would represent the transition from Middle to Late Stone Age. The hard spherical limestone concretion was within a fossil soil and ricocheted off the pick blade at considerable speed. I felt a terrific blow to my eye and saw a flash of light. I called down quietly to Stan, who was standing at the base of the cliff, that I had lost the use of my left eye and climbed carefully down to retrieve my pack and field gear. Five hours later, in Nairobi Hospital, I was chatting to ophthalmologist and consultant eye specialist Dr Fayaz Khan, who had been trained in part by the great New Zealand eye surgeon Fred Hollows, as he carefully removed the fragments of grit that were embedded in the eyeball. Thanks to the delights of pethidine, this was a painless procedure. Luckily, the damage was not too severe and with prescription glasses I could see perfectly well, although the pupil remained dilated. A few weeks later I was in Inner Mongolia (see Chap. 17), armed with plenty of eye drops to keep the surface of the eye moist.

During three further seasons in Kenya (2001, 2002, 2003) we continued our work in the western rift and extended it as far as Lake Naivasha and Lake Elmenteita. We were now finding evidence of significant changes at the transition from Middle to Late Stone Age about forty five thousand years ago. Ostrich egg-shell beads appeared and the quality of obsidian used to fashion stone tools also increased as the Late Stone Age peoples began to search out sources of high quality obsidian from much further away.

Our nineteen-year-old daughter Rebecca joined us in 2003 and quickly mastered some appropriate Swahili proverbs. One morning she was working down by a river examining samples of volcanic ash under the microscope when I paid her a visit and noticed that a large male baboon was watching her with considerable interest. I pointed this out to her very gently and she swivelled round. 'What should I do?' she asked rather plaintively. 'Carry on working' was my reply. On first arriving at our camp in Maasai country up on a flat grassy site flanked on one side by hills and on another by steep cliffs leading down to the river she could hear the hyenas cackling and the occasional distant roar of a hungry lion. I indicated that she should pitch her tent and get to sleep as it was close to midnight when we arrived at the camp. 'Where are the Maasai warriors who are meant to protect me?' she asked somewhat apprehensively. 'All around you!' I replied airily, 'but you can't see them.' They were, of course, asleep in their tents. Next day our friend Nepatao ole Simpai, a Maasai sub-chief, took her under his wing and said he would look after her. This was reassuring because there was a solitary old bull buffalo, which used to doze down by the river. One morning while we were all eating breakfast Stan asked me: 'did you hear the buffalo in the night? It went right past my tent!' 'What time was that, Stan?' I asked innocently. 'About five!' he replied. 'That's odd' I said, 'I was walking past your tent about five this morning and did not see any buffalo!' We had no more buffalo scares after that. Still, wild animals can wander through camps at night. Louis Leakey advised Bill Bishop during his first visit to Olduvai to move his camp bed to one side. 'The rhino walks past that spot every night'. Bill decided this was a joke and decided to ignore the advice. Next morning Louis asked him if he had heard the rhino go past. 'Not a thing' said Bill, at which Louis pointed out a great steaming heap of warm rhino dung just by his bed. Bill told me that tale himself when we first met at a Quaternary geology conference in New Zealand in 1973.

Simpai had some impressive scars on the front of both shoulders. I asked him how he got them. He said he had been asleep in a hut in his manyatta when a herd boy rushed in to tell him that a lion was attacking one of his cows. He grabbed his spear and ran towards the lion, which let go of the cow, roared and charged. Sempai threw his spear and killed the lion. The lion collapsed on top of Sempai with its claws embedded in his shoulders. When I asked him if he was at all scared when the lion charged he looked puzzled. 'Scared? No. I was very angry. It was my favourite cow!'

Near one of our camps close to Lake Naivasha, the landowner had erected electrified fences to keep buffaloes from damaging his crops. One day we needed to cross the fence so I checked to see if there was any current in the top three strands. There was none. Reassured, I bent down and lifted one leg across the bottom wire, which proved to be electrified. My Kamba workers mimed the scene back in camp that evening, and concluded very poetically: 'at that point, Martin began to speak in the language of his forefathers!'

My graduate student Peter Glasby proved a great asset during our last two seasons in Kenya. Stan discovered early one morning that he had no hydrochloric acid in his bottle with which to test soils for the presence of calcium carbonate. Peter offered to go and make some, disappeared, and returned a few minutes later with the required acid in a stout plastic bag. He had used battery acid (sulphuric acid) mixed with salt. On another occasion Peter made a rocket from paper, wire and cotton. It was a clear night and the rocket, propelled by hot air generated by a tray of burning cotton, rose vertically for some hundreds of metres, glowing in the night, and then blew in the direction of Lake Victoria to the west. Simpai commented that the Maasai laibon (diviner and healer) would receive visits next day from many puzzled and anxious warriors! On another occasion bees had stung the wife of one of our local men on the face, which had swollen so that she could not see. Peter cut open an onion from her garden and rubbed it gently across the areas stung. The treatment worked and as the pain and swelling subsided, his prestige grew!

One of our more unusual sites was a long and narrow cave known in Kikuyu as Ngomot Ngai (breath of God). We used ropes to descend into the entrance and then proceeded for several hundred metres laterally through a succession of water-lain tuffs and fossil soils with a total vertical thickness of about twenty metres. We sampled each unit and brought the samples to the surface for a more thorough description. The cave was well ventilated but we finally reached a narrow crack only about 10 cm wide and had to end our mapping. The cave was full of bats, which deterred some of our party from entering the cave. We concluded that the cave had probably formed as a result of a small earthquake opening up an underground fissure. Water flowing down into the cave from above completed the job of excavation during the wet season.

Chapter 20
Mauritania, France, Argentina (2004–2014)

Conferences and workshops play a very important role in the life of a natural scientist. They may range from gargantuan international meetings with many thousands of participants to highly focussed small groups of a few tens of specialists meeting to discuss a particular problem. Such conferences often serve as refresher courses exchanging the latest results in a particular scientific discipline but an equally important role is the development of an informal network and the exchange of ideas on a one to one basis. It is all part of belonging to the international community of scholars. Two of the very best conferences in which I have taken part in recent years were in Mauritania in January 2004 and in Argentina in November 2014, and I shall try to explain why I found them so worthwhile.

Professor Suzanne Leroy from Brunel University in London organised the conference to Mauretania (Figs. 20.1, 20.2, 20.3 and 20.4). Suzanne is a specialist in the reconstruction of past changes in plant cover using fossil pollen grains. She had secured funding from two international scientific bodies to support studies of extreme events and their impact on human societies and had resolved that our discussions would take place during our hike across the rugged plateaux, the beds of former lakes, and the great sand dunes of the Mauritanian desert, without benefit of electricity or slide projectors. This was near the Majâbat Al-Koubrâ or 'Empty Quarter' of Mauritania where the great Saharan naturalist and polymath Théodore Monod had carried out his explorations by camel in the 1950s. One of our camels carried a small blackboard and trestle, so chalk and talk was the order of the day. The landscape was spectacular, with the huge meteorite crater at Aouelloul (Fig. 20.5) one of the highlights. Our walk across the desert ended at the oasis of El

M. Williams, *Nile Waters, Saharan Sands*, Springer Biographies,
DOI 10.1007/978-3-319-25445-6_20

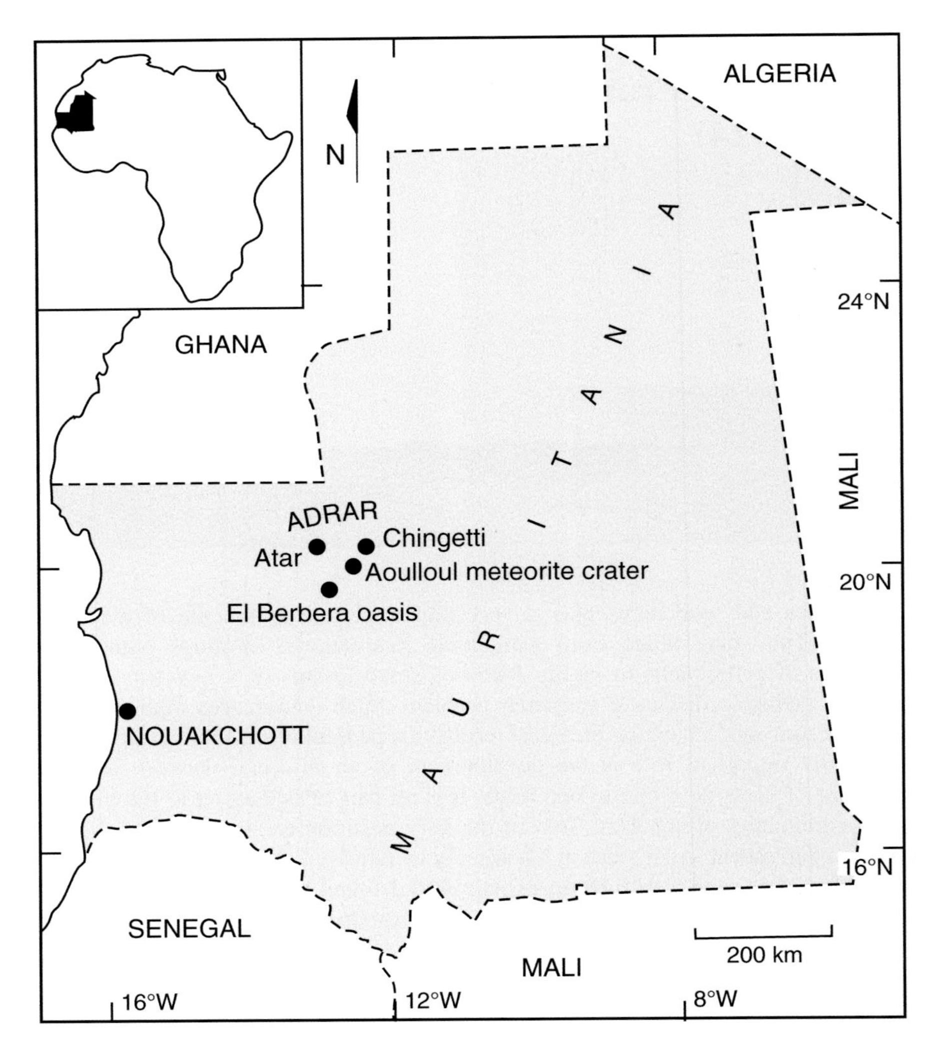

Fig. 20.1 Mauritania (2004)

Berbera, where we rediscovered the childlike joys of tobogganing down a steep sand slide (Fig. 20.4). We celebrated the end of our walk with songs and an impromptu dance led by my old friend Mohammed Ben Azzouz (Fig. 20.2) (see Chap. 12), while I tapped out a suitable rhythm on a small local drum (Fig. 20.3).

Pascal Lluch and Damien Parisse from the French adventure tour company Homme et Montagnes saw to all our travel arrangements, food and local guides and did so with superb efficiency and minimum fuss. Both were experienced desert and mountain guides. Pascal was keen on the local natural history and knew the Sahara well. He was great company and later stayed with us at our home in Adelaide

Fig. 20.2 Suzanne Leroy and Mohamed Ben Azzouz celebrating the end of our walk, Mauritania, January 2004. *Photo* Dr Hélène Jousse

Fig. 20.3 Françoise Gasse and the author celebrating the end of the desert walk, Mauritania, January 2004. *Photo* Dr Hélène Jousse

during a trip he had arranged across Australia. I stayed with Damien later that year at his house high in the French Alps, while he was checking on the feeding habits of wolves, which had arrived recently from Italy. He and his partner were shepherds and she made one of the tastiest cheeses I have ever eaten.

One of the topics we explored on this trip was the question of past changes in climate and the role of humans in causing land degradation and desertification. The alternation of extreme floods and droughts so characteristic of deserts and so vividly portrayed by the Old Testament scribes has attracted the attention of local scholars for many hundreds of years. For China there is an invaluable compilation of wet

Fig. 20.4 Author discovers the fun of sand slides, Mauritania, January 2004. *Photo* Dr Hélène Jousse

Fig. 20.5 On the rim of the meteorite crater at Aouelloul, Mauritania, January 2004. *Photo* Dr Hélène Jousse

and dry spells extending back five hundred years. North Africa has chronicles of famines and droughts extending back to at least 1500 AD. One as yet untapped repository of such information is in the ancient library at Chinguetti, a small town with a turbulent past, located amidst the dunes of the Adrar region in central Mauritania. I had the good fortune in January 2004 to visit the house in which the ancient scrolls were stored. The librarian showed me rolled parchments made of vellum, some he considered to be up to a thousand years old. He explained that during past raids on the town, the scrolls would be hurriedly loaded into camel bags and sent out into the desert for safekeeping. The topics detailed in the scrolls included botany, medicine, astronomy, mathematics and climatic history. They remain an untapped archive for future scholars to investigate.

Scattered between the dunes there were abundant remains of Neolithic occupation sites, often associated with freshwater snail shells indicative of former ponds and small lakes in the hollows between the dunes. They were similar to the Neolithic sites that Dr Hélène Jousse, who was also with us, had been working on in the desert of Mali. Out on the gravel plains sporadic stone hand-axes and cleavers bore witness to an even earlier human presence during Early Stone Age times nearly half a million years ago. I remember pointing them out to the distinguished

Fig. 20.6 Stein-Erik Lauritzen and the author trudging through the dunes, Mauritania, January 2004. *Photo* Dr Hélène Jousse

Norwegian cave scientist Stein-Erik Lauritzen (Fig. 20.6) from the University of Bergen (where my close friend Mike Talbot worked) as we walked over them. A day later, after a long walk across the dunes, we experienced torrential rain and for the first time had to sleep under canvas (Fig. 20.7a–c). The moist air blowing eastwards from the Atlantic to release its surplus moisture many hundreds of kilometres inland made it easier for us to visualize the reality of wetter conditions in the heart of the Mauritanian desert during the Neolithic.

My friend and colleague Dr Françoise Gasse (Fig. 20.3) (see Chap. 11) was with us on the desert leg of this conference. We had ample leisure to discuss plans for future joint research. We were both keen to attempt a detailed reconstruction of the climatic fluctuations in the Southern Hemisphere during the last thirty thousand years, taking into account the evidence from both the oceans and the land. To make the job manageable, we decided to focus first on Africa south of the equator, before moving on to Australia and South America. Accordingly, I spent some months working at the *Centre Européen de Recherche et d'Enseignement des Géosciences de L'Environnement* or CEREGE (European Centre for Earth Science Environmental Research and Teaching), near Aix-en-Provence, with a team of French colleagues during 2005, 2006 and 2007. These research visits were an unforeseen by-product of the field conference in Mauritania. During my time there we stayed at Calas, a village situated amidst magnificent limestone scenery. We sometimes climbed Mont Saint Victoire (Figs. 20.8, 20.9a–c, 20.10 and 20.11), immortalised by Cézanne in his painting of that mountain.

The other conference I mentioned earlier was the Fourth Southern Deserts Conference held near Mendoza in central Argentina during November 2014. The prime mover and chief organiser was Dr Ramiro Barbarena, a brilliant young archaeologist at the University of Mendoza. Many researchers and graduate students from Argentina and Chile as well as smaller numbers of researchers from

Fig. 20.7 Rain clouds over the desert, Mauritania, January 2004. *Photos* Dr Hélène Jousse

Australia, Europe and the USA took part in the conference. The field excursion among the rivers, lakes and volcanoes of northern Patagonia took us into some of the most stunning scenery I had ever had the privilege to visit, with the ever-present vista of the snow-capped Andes (Figs. 20.11 and 20.12) to the west, small herds of guanacos perched on top of low cinder cones to cool themselves in the breeze, and mighty condors with their three-metre wing-span gliding in the thermals overhead. A few months before coming to this conference, Cambridge University Press in New York had published a lengthy book I had written called *Climate Change in*

Fig. 20.8 Base of Mont Saint Victoire, Provence, France, May 2012. *Photo* Frances Williams

Deserts: Past, Present and Future. One of the chapters was on the deserts of South America. After attending this conference, chatting to my colleagues from Argentina and Chile, and admiring the posters prepared by the graduate students and their supervisors, I began to realise all too clearly just how little I really knew and how much more there was for me to discover. Michelangelo famously remarked when in his seventies he was asked what he was up to painting frescoes on the ceiling of the Sistine Chapel in Rome: '*Ancora imparo*' ('I am still learning'). I have always thought this to be a worthy motto for any aspiring student of planet earth.

Other earlier conferences that also proved illuminating and memorable were ones held in South Africa in mid-1979 and in Israel in 1988 and 2009. In each case the proportion of formal talks was kept to a decent minimum to allow time for well-organised and leisurely field excursions. In South Africa this allowed ample time for visits to the coastal prehistoric site of Klasies River Mouth where Hilary and Janette Deacon had been excavating and to Boomplaas Cave in the southern Cape, which showed early evidence of sheep herding and some spectacular rock paintings and engravings.

The 1988 field excursion to the southern Negev Desert in Israel was organised by Professors Asher Schick and Ran Gerson, and showed how quantitative data about erosion in this arid region could be obtained using quite simple monitoring techniques. Given my own work on erosion rates in Australia, I found these experiments especially interesting. I returned to Israel in late 2009 for another mainly field-based conference run by Professors Yehouda Enzel and Moti Stein, two ebullient and very sharp earth scientists. The scenery was as spectacular as the quality of the work we were shown. I stayed on after the conference and had the privilege of being shown first hand the dune sites in the southern Negev where Professor Aaron Yair was monitoring the effects of plant crusts on runoff and

Fig. 20.9 Plants along the slopes of Mont Saint Victoire, Provence, France, May 2012. *Photos* Frances Williams

Fig. 20.10 Rebecca on the summit of Mont Saint Victoire, Provence, France, May 2012. *Photo* Frances Williams

Fig. 20.11 Glaciated peaks of the Andes between Santiago, Chile and Mendoza, Argentina, November 2014. *Photo* Frances Williams

Fig. 20.12 Glacial cirques in the high Andes between Santiago, Chile and Mendoza, Argentina, November 2014. *Photo* Frances Williams

infiltration. For me the highlight was staying with Dr Yoav Avni and his family in the southern Negev and having Yoav explain the geological evolution of that extraordinary landscape which he had mapped and knew so well. Yoav had spent six months on study leave in my department at Adelaide the previous year, and had carried out a valuable reconnaissance survey of gully erosion in the arid Flinders Ranges of South Australia.

Chapter 21
Back to the Nile (2005–2012)

There is a well-known Arabic saying that who has once drunk the waters of the Nile must needs return. I have certainly drunk more than my fair share of Nile water and so am ever alert to the siren call of that noble river. Consequently, when my Scottish friend and soil surveyor colleague Neil Munro sent me an invitation to join him in early 2005 for some fieldwork on either side of the Main Nile in northern Sudan (Fig. 21.1), I did not hesitate for long (Figs. 21.2, 21.2. 21.3, 21.4, 21.5 and 21.6). Neil and a team of experienced Sudanese soil surveyors led by Professor Osman et Tom, head of the Soil Survey Division at the Gezira Research Station in Wad Medani (see Chap. 5) had been mapping the soils over a vast area of northern Sudan as part of the Merowe Dam project. This dam was built on the fourth cataract near the town of Merowe between 2003 and 2009 and has been a source of enormous hardship for local communities displaced by the rising waters of the

M. Williams, *Nile Waters, Saharan Sands*, Springer Biographies,
DOI 10.1007/978-3-319-25445-6_21

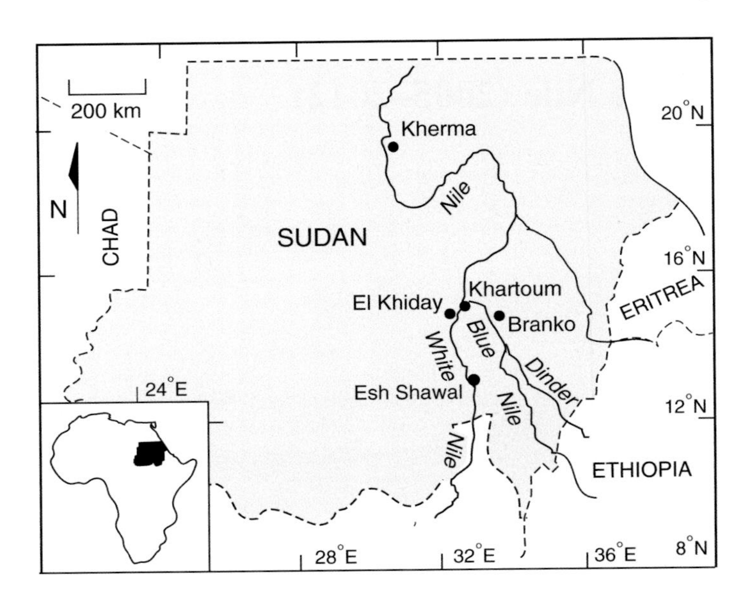

Fig. 21.1 Nile Valley, Sudan (2005–2012)

Fig. 21.2 Osman and author collecting samples for dating times of dune activity, northern Sudan, January 2010. *Photo* Neil Munro

reservoir. Nor was a rigorous environmental impact assessment ever carried out. Time will tell whether the costs outweigh the benefits.

Fig. 21.3 Desert west of Nile, northern Sudan, January 2005

Fig. 21.4 Nubian Sandstone hills west of Nile, northern Sudan, January 2005

Neil was pinned down by a nasty attack of malaria when I arrived and had not eaten for a while. He revived when I made it clear that I did intend to eat and that the sooner we got into the field the better. By Burns Night Neil had recovered sufficiently that he donned his kilt and sporran and ceremoniously despatched a haggis while reciting the immortal ode to a haggis penned by Robert Burns.

Neil's soil pits were scattered across the landscape. Neil, Osman (who became an esteemed friend and field companion on a number of later visits to the Nile valley) and I revisited these sites and I logged and sampled them for radiocarbon and optical dating (Fig. 21.2). There were good practical reasons for trying to obtain

Fig. 21.5 Fossil Nile perch skeleton in the desert west of the Nile, northern Sudan, January 2005

Fig. 21.6 Jemal, local camel herder, Osman, Neil Munro, desert west of the Nile, northern Sudan, January 2005

reliable ages. Once we knew the ages of the various sediments we could work out when and how the different soils had formed and better understand the reasons for their varying physical and chemical properties. The landscape further west was vastly more spectacular than the arid plains we had been sampling, with eroded hills of Nubian Sandstone mantled in a patina of black desert varnish emerging from the dunes (Figs. 22.3 and 22.4).

At one site we had a spot of bother from an enraged elderly local farmer who came hurling rocks at Neil, whom he accused of trying to steal his land. I walked up to the angry old man, showed him the snail shells I wanted to collect and engaged him in conversation. I noticed that he was missing a few teeth and asked if the strains of supporting his wives had loosened his teeth, and extended my sympathy, which seemed to cheer him up. We then chatted about his family, of whom he was very proud, with two sons training to be doctors at the University of Khartoum. Later, out in the open desert west of the Nile, a local Kababish camel herder brought us to a site where he had built an enclosure of rocks to protect the fossil remains of a large Nile perch (Figs. 20.5 and 20.6). We also investigated some pits dug into the floor of an ancient lake. This lake had been fed from an overflow channel of the Nile when it had flooded to a higher level than today between ten and seven thousand years ago, when the climate was significantly wetter throughout northern Sudan.

I thought it time now for a concerted study of Nile floods across the entire basin. Accordingly, I gathered a first rate international research team together and obtained funding from the Australian Research Council. In putting together a team to work in remote, difficult and sometimes dangerous places, I have always borne in mind three essential qualities. First is a high degree of professional competence allied to a willingness to think beyond one's immediate professional discipline. Next are self-reliance, initiative and fortitude, so that one does not become a burden on other team members. Finally, I value thoughtfulness towards others without infringing on their need for privacy.

Our initial focus was on floods and droughts during the last thirty thousand years but I later extended it to cover the last 125,000 years, or a full glacial-interglacial cycle. Members of the team included Mike Talbot, with whom I had worked in Niger and Australia, Geoff Duller (optical dating) and Mark Macklin (river floods), both from Aberystwyth University in north Wales, and Jamie Woodward (geo-archaeology and river floods) from the University of Manchester. Mark and Jamie concentrated on northern Sudan and worked closely with Derek Welsby from the British Museum, who had been excavating archaeological sites in that region for many years. This project has been highly successful, has greatly enlarged our understanding of the recent geological history of the Nile, and has led to many scientific publications.

The following year (2006) I decided to focus my efforts on trying to determine when the last major flooding of the lower White Nile valley had taken place. Don Adamson and I had earlier mapped the shoreline of a huge lake that had flooded the lower White Nile valley to an elevation of 396 m above sea level. We had used colour satellite images to do this, together with detailed topographic maps with contours every 50 cm. The eastern shorelines of this lake formed giant curves or cusps ranging from ten to thirty kilometres from tip to tip, and the lake itself was up to 80 km wide and over 500 km long. The curved beaches were tethered to rocky headlands, usually rounded granite hills of ancient rocks protruding through the thin mantle of younger sediments. Osman, my PhD student David Haberlah (who spoke good Arabic and also proved an ace navigator), geologist Dr Tim Barrows (a specialist in what is called cosmogenic nuclide dating) and I resolved to date the shoreline directly. To do this we acquired the services of a friendly backhoe driver and excavated a deep trench through the sands and gravels of the 386 m beach. The age we obtained was 110,000 years, and was a time of much wetter climate in this region, with important implications for prehistoric human migration routes to the north and up the Nile valley at this time.

I had long suspected that changes in the climate and vegetation cover in the Ethiopian highlands would have had a major influence on the Blue Nile and the Atbara rivers. Accordingly, in 2009 we began to investigate the history of recent glaciation in the Semien highlands of Ethiopia, close to the headwaters of both the Tekezze-Atbara and Blue Nile rivers. We (Frances, Tim Barrows and I) camped at a place called Chenek among some of the most spectacular scenery in Africa, at an elevation of 3600 m, with troops of Gelada baboons grazing peacefully around our tents while graceful lammergeiers cruised in the thermals above. This was the land of the Walia ibex and the Semien wolf—a land of deep gorges and great precipices, of ancient eroded volcanoes and vast horizontal basalt flows. Working above 4000 m, we located and sampled the glacial moraines flanking two high mountains. The sampling involved the use of a cold chisel and a sledgehammer to collect fist-sized chunks of rock from the moraine boulders. Our aim was to date the moraines by measuring the amount of cosmogenic nuclides trapped within the rock, a complex technique of which Dr Tim Barrows was an acknowledged expert.

So far so good, but what I was still lacking was a detailed history of major floods and droughts in the lower Blue and White Nile valleys to compare with the indirect record of past floods preserved in marine sediments on the floor of the eastern Mediterranean. During two field seasons in the first few months of 2010 and 2011 we mapped, logged and sampled a number of key sites. These sites included old channels of the Blue Nile, sand dunes flanked by flood plain clays, and bank sections along the Dinder River (a major Blue Nile tributary) within sight of the dark cliffs of the Ethiopian escarpment. At various times, our team consisted of Neil, Osman, Professor Yasin el Hejaz (a hydro-geologist and an old friend with whom Don Adamson and I had worked at Esh Shawal in 1983), Professor Geoff Duller—a world authority on optical dating, and myself.

There were occasional moments of drama to enliven the day. At the fossil-bearing site of Branko immediately east of the present Blue Nile we were inspecting some fossilised tree trunks when an enraged peasant farmer came charging down towards us flourishing a shovel and threatening to decapitate our young local guide. The angry man was terrified that the government would lay claim to his land because there were fossils nearby. I sought to mollify him by telling him that the fossils were of no great interest and that the government had far better things to worry about and offered him a new shovel to replace his old one, which was broken. We then adjourned for lunch in the village and good humour was restored. I noticed that our young guide was still rattled so I asked him: 'Endek wahed ras wella nus?' ('Do you still have a complete head or only half?'). This provoked a laugh, and he relaxed and began to eat.

On another occasion I was inside a trench on top of a sand dune when an excited elderly man came galloping up on his donkey waving an axe. I climbed out of the trench and greeted him in colloquial Sudanese Arabic, playing on words. 'Inta shuf hashab, wella inta shuf harami?' ('Are you seeking wood or thieves?'). To which he replied: 'Wallahi, kuhullu Sudani harami!' ('All Sudanese are thieves!'). I looked suitably pained and replied: 'Lakin ana nus Sudani! Ana shinu eh?' ('But I'm half Sudanese, so what does that make me?') He was the guardian of a grove of neem trees and of the 'toothbrush tree' *Salvadora persica*. The neem tree (*Azadirachta indica*) comes from India, is a member of the mahogany family, and its leaves secrete an oil that is valued as a mosquito deterrent. Keen to get back to my work, I suggested to the old man that he should go and tend his toothbrush trees while he still had some teeth in his head to enjoy them. Cackling with delight, he rode away waving his axe. There were many such encounters, prompted more by curiosity than by anything else, and often culminating in an invitation to share a meal. I always felt at home among these good-humoured, courteous and hospitable folk and certainly never felt in any way unsafe or threatened. Knowing the language and being able to tell stories has helped enormously. Wherever I worked in Africa I have always tried (not always very successfully) to master a basic working vocabulary but knowing Arabic proved helpful in a number of countries besides the Sudan.

My latest encounters with the Nile arose from invitations to work with two teams of archaeologists, one Italian and one Swiss. Dr Donatella Usai and her husband Dr Sandro Salvatori had been working for some years excavating a series of Mesolithic and Neolithic sites west of the lower White Nile near the village of El Khiday. I joined their team in November 2011 and set about collecting samples for optical and radiocarbon dating, including at the site of a vast former wetland. The interest of these sites was twofold. The Mesolithic was the precursor to the Neolithic, during which plant and animal domestication led to a revolution in lifestyle, with the cultivation of cereal crops and the herding of cattle, sheep and goats replacing the life of the nomadic hunter-fisher-gatherers. All previously excavated Mesolithic sites in central Sudan had been disturbed to some degree by later human occupation so that it was not possible to be certain that some parts of the site were still in primary context. Through trial and error, Donatella and Sandro had hit upon a series of ways of being able to recognise undisturbed, primary context sites, and so had added immensely to our knowledge of the complexity of the Mesolithic. This was a first for Sudanese archaeology.

In January 2012, I joined the Swiss team led by Professor Matthieu Honegger from the University of Neuchâtel, who were excavating a large number of prehistoric burial sites in a huge cemetery near the town of Kerma east of the Main Nile in northern Sudan. One especially interesting site called the Wadi el Arab site was out in the open desert to the east in the lee of an outcrop of hard Nubian Sandstone. We were plagued from time to time at this site by the nocturnal digging activities of grave robbers, who were convinced that we were in search of gold. We found their trenches dug into the sterile sandstone hill slopes where they had no hope of ever finding gold.

I later enjoyed the hospitality of the king of Argo Island on the Nile and later met the Sufi mystics (sometimes known as the 'dancing dervishes') outside their mosque in Omdurman (and close to a large cemetery) during their Friday late afternoon ceremonies culminating in some of them spinning like tops on one leg. Just before the sun went down all the onlookers vanished. I was wondering why when my driver Jemal ushered me away in something of a panic. These Sufi mystics are reputed to be able to raise the dead once the sun has set and nobody wanted to be there to witness that event.

The final outcome of all the Nile valley work we carried out in field and laboratory between 2005 and 2012 was very pleasing. For the first time we had a record of major episodes of Nile flooding extending back over 125,000 years, which we could now match with the records obtained from marine sediment cores in the eastern Mediterranean. Furthermore, we could relate these flooding episodes to changes in the strength of the African monsoon associated with changes in the tilt of the earth's axis and with changes in the distance of the earth from the sun during that long interval of time.

The work with the Italian and Swiss teams also brought new insights into the prehistory of the Nile valley during the last fifteen thousand years. We now had a well-dated record of wet and dry phases in the Nile valley, which we could relate to changes in the movements and activities of prehistoric communities in a way that had not previously been possible. The Nile had amply repaid the attention we had paid her.

Chapter 22
Epilogue

The time has come to take stock of what we have actually achieved during the past fifty years. I say 'we' advisedly, because much of my work was in cooperation with others. So, what new insights did I and my field companions gain about the earth we live on, and what are the big questions we were asking that still remain unresolved?

After the two expeditions to the southeast Libyan Desert in the northern summers of 1962 and 1963, I had become fully convinced that the climate in this vast region had been significantly wetter on a number of occasions in the past. I did not know when it had been wetter and I certainly did not know why. However, I did ask myself whether the desiccation of the Sahara was caused by long-term changes in

M. Williams, *Nile Waters, Saharan Sands*, Springer Biographies,
DOI 10.1007/978-3-319-25445-6_22

climate unrelated to any human activities, or whether prehistoric humans may have played a role through overgrazing by herds of Neolithic cattle. I was inclined to doubt this latter possibility because it could not explain repeated changes of climate, well before the advent of the Neolithic, for which there was good evidence. I also wondered what had been happening in other parts of the arid and semi-arid world. Had they too once been wetter and more hospitable for humans than they were today?

My early days of soil surveys in the lower Blue and White Nile valleys of central Sudan during 1962–1964 had convinced me that a thorough knowledge of the depositional history of the Nile and its tributaries was a prerequisite to understanding the origins and characteristics of the soils in that vast region. The only way to achieve such an understanding was to embark upon a systematic programme of dating the alluvial sediments laid down by the Blue and White Nile rivers, with particular attention to mapping the elevations attained by past floods. This work led in due course to a reasonably detailed history of past floods and droughts spanning the last fifteen thousand years, related to changes in the intensity of the summer monsoon. I was eventually able to compare this Nile flood history with events in the central Sahara, Ethiopia, India and China, all of which reflected changes in the summer monsoons. We were ultimately able to extend the Nile flood history back to 125,000 years ago with some confidence. By now it had become clear that changes in the intensity of the African and Asian summer monsoons were controlled by changes in the amount of heat received at the surface of the earth, especially in the inter-tropical zone. Such changes were caused by subtle cyclical changes in the tilt of the earth's axis (controlling seasonality) and in the distance of the earth from the sun (controlling the total amount of heat received from the sun in the upper atmosphere).

By 1980 we had published our initial record of high and low flow episodes in the Blue and White Nile covering the past fifteen thousand years, which we refined in subsequent years. Don Adamson and I then decided that it was time for us to compare the *historic* record of annual Nile floods with the available data for floods and droughts in India, China, Australia, and Indonesia. To our surprise, and gratification, we found a recurrent pattern of synchronous extreme floods (and extreme droughts) in each of these regions associated with changes in sea surface temperatures in the equatorial Pacific. In short, El Niño events off the coast of Peru were harbingers of drought in eastern Australia, eastern China, Java, India and the Nile basin, while La Niña events in Peru led to widespread floods in those same regions. (During El Niño years the waters off Peru become warmer than average and during La Niña years the sea surface off Peru becomes colder than average). Armed with this knowledge, I took a hard look at a popular and widely accepted theory relating droughts in the Sahel region of Africa to overgrazing and a change in the surface albedo or reflection of incoming solar radiation, and found it wanting. The hardship caused by drought is bad enough without blaming humans for having caused them! Since that time our capacity to predict future droughts (and floods) based on monitoring changes in sea surface temperatures in selected parts of the oceans has enabled governments and farmers around the world to plan ahead and take appropriate adaptive or preventative measures.

My interest in soil erosion, land degradation and desertification had not waned since my doctoral research on rates of hill slope erosion in tropical northern and temperate southeast Australia in the mid-1960s. Having chaired the Natural Resources Council of South Australia during 1994–1998, I was acutely aware that good government environmental policy is impossible without a solid foundation of good science.

In 1993 I was asked by the respective directors of the United Nations Environment Programme (UNEP) and the World Meteorological Organisation (WMO) to prepare a major report on the interactions between desertification and climate as a precursor to the drafting of the UN Convention to Combat Desertification. I suggested that climatologist Professor Bob Balling from the State University of Arizona at Tempe should be my co-author, and this suggestion was accepted. Preparing this report was an excellent opportunity to review a great deal of research from around the world and to summarise some of my own work. My feelings in writing this relatively factual report are best expressed by quoting Winston Churchill in *The Story of the Malakand Field Force* (1898): '*I pass with relief from the tossing sea of Cause and Theory to the firm ground of Result and Fact*'. Nevertheless, in any science there will always be a role for speculation and hypothesis building, as Galileo's brilliant but ill-fated contemporary Giordano Bruno (1548–1600) reminds us in *De gli heroici furore* (1585): '*Se non è vero, è molto ben trovato: se non è cossi, è molto bene iscusato l'uno per l'altro*' (put briefly: 'It may not be true, but it is well contrived').

Distinguishing between human influences and natural causes of desertification and land degradation remains a perennial problem. There are no easy answers and each case needs to be assessed in its own right. Certainly, there are many instances where attempts to install soil conservation measures are neither useful nor necessary. We still can learn from the Romans who turned the processes of hill slope erosion in the coastal valleys of North Africa to their advantage and trapped enough silt behind porous stone dams along the valley floors to enable the local farmers to grow wheat and barley, date palms and olives, just as they do today in the Matmata Hills of Tunisia.

An important theme running through my work has been a desire to reconstruct the environments associated with prehistoric human occupation, often in areas now too arid and too inhospitable for any permanent human settlement. In early 1970, working closely with Desmond Clark at Adrar Bous, an isolated mountain in the geographical heart of the Sahara, we were able to identify a series of stages in prehistoric occupation coinciding with wetter climatic phases extending back from Neolithic to Early Stone Age. At that time we had no clear idea of the age of these cultural episodes nor of the associated climatic fluctuations—all of that knowledge came later, as new techniques of dating rocks and sediments were developed and perfected.

As far as my own studies of the geology, soils and landforms of Australia are concerned, I have but scratched the surface. A task for the future is to integrate our

Fig. 22.1 Patrick Williams bouldering in the Adelaide hills, South Australia, 2013. *Photo* Allyson Williams

studies of landscape evolution more thoroughly than we have yet achieved with our understanding of the evolution of the Australian flora and fauna, much of it now based upon insights from molecular biology but still lacking a firm and independent chronometric framework. If the magnitude of this task seems daunting, we can find consolation in some words penned in another context altogether by Charles Darwin in *The Descent of Man* (1871): '*…the discovery of fossil remains has been a very slow and fortuitous process. Nor should it be forgotten that those regions which are the most likely to afford remains connecting man with some extinct ape-like creature, have not as yet been searched by geologists*'. There are many parts of the vast continent of Australia that have not as yet been adequately investigated by students of landscape development. Of course, this applies equally well to Africa, Asia and South America, and, indeed, elsewhere.

Throughout my work I have always believed that a proper understanding of the present is not possible without a thorough knowledge of past events. Or, as the great poet T.S. Eliot (1888–1965) put it so well:

Time present and time past
Are both perhaps present in time future,
And time future contained in time past…
Four Quartets (1944), 'Burnt Norton'

Finally, if asked what advice I might offer future enquirers into landscape history, I can do no better than quote the words of the brilliant French airman Georges Guynemer (1894–1917): '*Atteindre au delà, chercher à s'élever toujours*' ('Strive further, seek always to go higher'). Rock climbers like my son Patrick Williams who practice the noble art of bouldering (Fig. 22.1) will know exactly what I mean.

Further Reading

This bibliography provides access to specialist details of matters discussed in the individual chapters.

Background

Adamson, D.A., Williams, M.A.J. and Baxter, J.T. (1987). Complex late Quaternary alluvial history in the Nile, Murray-Darling and Ganges basins: Three river systems presently linked to the Southern Oscillation. In: V. Gardiner (ed.), Proceedings of the First International Conference on Geomorphology, Manchester, September 1985, John Wiley & Sons, Chichester, Part II, 875-887.

Williams, M.A.J. (1985). On becoming human: geographical background to cultural evolution. 11th Griffith Taylor Memorial Lecture. Australian Geographer 17, 175-185.

Williams, M.A.J. (1994). Cenozoic Climatic Changes in Deserts: A Synthesis. In: A.D. Abrahams and A.J. Parsons (eds) Geomorphology of Desert Environments. Chapman and Hall, New York, 644-670.

Williams, M. (2002). Deserts. In: M.C. MacCracken and J.S. Perry (eds), Encyclopedia of Global Environmental Change (ISBN-0-471-97796-9) Volume 1: The earth system: physical and chemical dimensions of global environmental change, pp 332-343. Wiley, Chichester.

Williams, M. (2002). Desertification, Definition of. In: I. Douglas (ed), Encyclopedia of Global Environmental Change (ISBN-0-471-97796-9) Volume 3: Causes and consequences of global environmental change, p. 282. Wiley, Chichester.

Williams, M.A.J. (2002). Desertification. In: I. Douglas (ed), Encyclopaedia of Global Environmental Change (ISBN 0-471-97796-9) Volume 3: Causes and consequences of global environmental change, pp 282-290. Wiley, Chichester.

Williams, M.A.J. (2002). Desertification Convention. In: Mostafa K. Tolba (ed), Encyclopaedia of Global Environmental Change (ISBN 0-471-97796-9) Volume 5: Responding to global environmental change, pp 183-186. Wiley, Chichester.

Williams, M. (2003). Changing land use and environmental fluctuations in the African savanna. In: Bassett, T.J. and Crummey, D., editors, African Savannas: Global narratives and local knowledge of environmental change. James Currey, Oxford, pp.31-52.

Williams, M.A.J. (2009). Cenozoic climates in deserts. In: A J.Parsons and A. D. Abrahams (eds.), Geomorphology of Desert Environments, 2nd edition. Springer: Berlin, pp.799-824.

Williams, M. (2014). Climate Change in Deserts: Past, Present and Future. Cambridge University Press, New York, 629 pp.

Williams, M. (2014). Why are deserts dry? Geography Review (UK), 28 (2), 34-37.

Williams, M.A.J. (2011). Environmental Change. Chapter 31 in: Gregory, K., Walling, D. and Goudie, A. (eds), Handbook of Geomorphology. Sage: London, pp. 535-554.

M. Williams, *Nile Waters, Saharan Sands*, Springer Biographies,
DOI 10.1007/978-3-319-25445-6

Williams, M. (2015). Interactions between fluvial and eolian geomorphic systems and processes: Examples from the Sahara and Australia. Catena Special Issue: Landforms and geomorphic processes in arid and semi-arid areas. Catena (2014), http://dx.doi.org/10.1016/j.catena.2014.09.015

Williams, M., Dunkerley, D., De Deckker, P., Kershaw, P. and Chappell, J. (1998). Quaternary Environments, 2nd Edition, Arnold, London, 1998, 329 pp.

Chapter 3. Sheffield and the Pennines

Williams, M.A.J. (1964). Glacial breaches and sub-glacial channels in south-western Ireland. Irish Geography 5, 83-95.

Chapter 4. Libyan Desert

Pesce, A. (1968). Gemini Space Photographs of Libya and Tibesti. Petroleum Exploration Society of Libya, Tripoli, 81 pp.

Williams, M.A.J. and Hall, D.N. (1965). Recent expeditions to Libya from the Royal Military Academy, Sandhurst. Geographical Journal 131, 482-501.

Chapter 5. Blue and White Nile Valleys, Sudan

Gaitskell, A. (1959). Gezira: A Story of Development in the Sudan. Faber and Faber, London, 372 pp.

Theobald, A.B. (1951). The Mahdīya: A History of the Anglo-Egyptian Sudan 1881-1899. Longmans, London, 273 pp.

Williams, M.A.J. (1966). Age of alluvial clays in the western Gezira, Republic of the Sudan. Nature 211, 270-1.

Williams, M.A.J. (1968). A dune catena on the clay plains of the west central Gezira, Republic of the Sudan. Journal of Soil Science 19(2), 367-378.

Williams, M.A.J. (1968). Soil salinity in the west central Gezira, Republic of the Sudan. Soil Science 105(6), 451-464.

Williams, M.A.J. (1968). The influence of salinity, alkalinity and clay content on the hydraulic conductivity of soils in the west central Gezira. African Soils 13(1), 35-60.

Chapter 6. Northern Territory, Australia

Davies, J.L. and Williams, M.A.J., eds (1978). Landform Evolution in Australasia. Australian National University Press, Canberra, 376 pp.

Galloway, R.W., Aldrick, J.M., Williams, M.A.J. and Story, R. (1976). Land systems of the Alligator Rivers area. In: R.Story et al. Lands of the Alligator Rivers Area, Northern Territory . CSIRO Land Research Series No. 38, 15-34. (CSIRO, Melbourne).

Haynes, C.D., Ridpath, M.G. and Williams, M.A.J., eds (1991). Monsoonal Australia: Landscape, ecology and man in the northern lowlands. Balkema, Rotterdam, 243 pp.

Williams, M.A.J. (1968). Termites and soil development near Brocks Creek, Northern Territory. Australian Journal of Science 31(4), 153-154

Williams, M.A.J. (1969). Prediction of rainsplash erosion in the seasonally wet tropics. Nature 222 (5195), 763-765.

Williams, M.A.J. (1969). Geology of the Adelaide-Alligator area. In: Lands of the Adelaide-Alligator area, Northern Territory, by R. Story et al. CSIRO Land Research Series, No. 25, 56-70. (Melbourne).
Williams, M.A.J. (1969). Geomorphology of the Adelaide-Alligator area. In: Lands of the Adelaide-Alligator area, Northern Territory, by R. Story et al. CSIRO land Research Series, No. 25, 71-94. (Melbourne).
Williams, M.A.J., Hooper, A.D.L. and Story, R. (1969). Land systems of the Adelaide-Alligator areas. In: Lands of the Adelaide-Alligator area, Northern Territory, by R. Story et al. CSIRO Land Research Series, No. 25, 25-48. (Melbourne).
Williams, M.A.J. (1973). The efficacy of creep and slopewash in tropical and temperate Australia. Australian Geographical Studies 11(1), 62-78.
Williams, M.A.J. (1974). Surface rock creep on sandstone slopes in the Northern Territory of Australia. Australian Geographer 12(5), 419-424.
Williams, M.A.J. (1976). Erosion in the Alligator Rivers area. In: R.Story et al. Lands of the Alligator Rivers area Northern Territory. CSIRO Land Research Series No. 38, 112-125. (Melbourne).
Williams, M.A.J. (1978). Termites, soils and landscape equilibrium in the Northern Territory of Australia. In: Landform Evolution in Australasia, ed. by J.L. Davies and M.A.J. Williams, 128-141 (Australian National University Press, Canberra).
Williams, M.A.J. (1978). Water as an eroding agent. In: K.M.W. Howes, editor, Studies of the Australian Arid Zone. III. Water in Rangelands, 79-89.
Clarke, M.F., Wasson, R.J. and Williams, M.A.J. (1979). Point Stuart Chenier and Holocene sea levels in Northern Australia. Search 10, 90-92.
Haynes, C.D., Ridpath, M.G. and Williams, M.A.J., eds (1991). Monsoonal Australia: Landscape, ecology and man in the northern lowlands. (Balkema, Rotterdam, 243 pp., including preface, index, editorial commentaries and glossary, vii-xii, 3-4, 39-40, 75-77, 195-6, 223-6, 229-31).
Williams, M.A.J. (1991). Evolution of the landscape. In: Haynes, C.D., Ridpath, M.G. and Williams, M.A.J., eds . Monsoonal Australia: Landscape, ecology and man in the northern lowlands. Balkema, Rotterdam, pp. 5-17.
Haynes, C.D., Ridpath, M.G. and Williams, M.A.J (1991). A torrid land. In: Monsoonal Australia, op. cit., 207-221.
Clarke, M.F., Williams, M.A.J. and Stokes, T. (1999). Soil creep: Problems raised by a 23 year study in Australia. Earth Surface Processes and Landforms 24, 151-175.

Chapter 7. Adrar Bous, Central Sahara

Williams, M.A.J. (1971) Geomorphology and Quaternary geology of Adrar Bous. Geographical Journal 137(4), 450-455.
Williams, M.A.J. (1973). Upper Quaternary sedimentation at Adrar Bous. In: J.D. Clark, M.A. J. Williams and A.B. Smith. The geomorphology and archaeology of Adrar Bous, Central Sahara: a preliminary report, 250-260. Quaternaria 17, 245-297.
Williams, M.A.J. (1975). Late Pleistocene tropical aridity synchronous in both hemispheres? Nature 253, 617-618.
Williams, M.A.J. (1976). Upper Quaternary stratigraphy of Adrar Bous, Republic of Niger, south central Sahara. Proceedings of the Seventh Pan-African Congress on Prehistory and Quaternary Studies, Addis Ababa, December 1971, 435-441.
Williams, M.A.J. (1976). Radiocarbon dating and late Quaternary Saharan climates: a discussion. Zeitschrift für Geomorphologie N.F. 20, 361-362.
Williams, M.A.J. (2008). Geology, geomorphology and prehistoric environments. In: D. Gifford-Gonzalez (editor) Adrar Bous: Archaeology of a Central Saharan Granitic Ring Complex in Niger. Royal Museum for Central Africa, Tervuren, 25-54.

Williams, M.A.J., Abell, P.I. and Sparks, B.W. (1987). Quaternary landforms, sediments, depositional environments and gastropod isotope ratios at Adrar Bous, Tenere Desert of Niger, south-central Sahara. In: L. Frostick and I. Reid (eds.) Desert Sediments: Ancient and Modern. Geological Society Special Publication No. 35, 105-125.

Williams, M., Glasby, P. and Blackwood, J. (2008). A note on an Acheulian biface from Adrar Bous, Tenere Desert, south central Sahara, Republic of Niger. Sahara 19, 85-90

Chapter 8. Ethiopian Highlands and Rift Valley

Abell, P.I. and Williams, M.A.J. (1989). Oxygen and carbon isotope ratios in gastropod shells as indicators of palaeoenvironments in the Afar region of Ethiopia. Palaeogeography, Palaeoclimatology, Palaeoecology 74, 265-278.

Assefa, G., Clark, J.D. and Williams, M.A.J. (1982). Late Cenozoic history and archaeology of the upper Webi Shebele basin, east-central Ethiopia. Sinet Ethiopian Journal of Science 5 (1), 27-46.

Barbetti, M., Clark, J.D., Williams, F.M. and Williams, M.A.J. (1980). Palaeomagnetism and the search for very ancient fireplaces in Africa: results from a million-year-old Acheulian site in Ethiopia. Anthropologie 18, 299-304.

Clark, J.D. and Williams, M.A.J. (1977). Recent archaeological research in southeastern Ethiopia (1974-75): Some Preliminary Results. Annales d'Ethiopie 11, 19-42.

Clark, J.D., Williams, M.A.J., Dakin, F., Gasse, F., Assefa, G., Bonnefille, R. and Adamson, D. (1979) Plio-Pleistocene environments in south central Ethiopia. Palaeoecology of Africa 11, 145-147.

Eberz, G.W., Williams, F.M. and Williams, M.A.J. (1988). Plio-Pleistocene volcanism and sedimentary facies changes at Gadeb prehistoric site, Ethiopia. Geologische Rundschau 77, 513-527.

Gasse, F., Richard, O., Robbe, D., Rognon, P. and Williams, M.A.J. (1980). Évolution tectonique et climatique de l'Afar central d'après les sédiments plio-pléistocènes. Bull. Soc. Géol. France (7) 22, 987-1001.

McDougall, I., Morton, W.H. and Williams, M.A.J. (1975). Age and rates of denudation of Trap Series basalts at Blue Nile gorge, Ethiopia. Nature 254, 207-209.

Williams, F. M., Williams, M.A.J. and Aumento, F. (2004). Tensional fissures and crustal extension rates in the northern part of the Main Ethiopian Rift. Journal of African Earth Sciences 38 (2), 183-197.

Williams, M.A.J. (1978). Palaeogeography of the Afar lakes: a critique. Palaeoecology of Africa 10, 183-4.

Williams, M.A.J. (1981). Recent tectonically-induced gully erosion at K'one, Metehara-Wolenchiti area, Ethiopian Rift. Sinet Ethiopian Journal of Science 4 (1), 1-11.

Williams, M.A.J. and Clark, J.D. (1976). Prehistory and Quaternary environments in southern Afar and on the Arussi-Harar plateau, Ethiopia. Palaeoecology of Africa 9, 98-100.

Williams, M.A.J., Bishop, P.M., Dakin, F.M. and Gillespie, R. (1977). Late Quaternary lake levels in southern Afar and the adjacent Ethiopian Rift. Nature 267, 690-693.

Williams, M.A.J., Street, F.A. and Dakin, F.M. (1978). Fossil periglacial deposits in the Semien Highlands, Ethiopia. Erdkunde 32, 40-46.

Williams, M.A.J. (1978). Palaeogeography of the Afar lakes: a critique. Palaeoecology of Africa 10, 183-4.

Williams, M.A.J., Williams, F.M., Gasse, F., Curtis, G.H. and Adamson, D.A. (1979) Plio-Pleistocene environments at Gadeb prehistoric site, Ethiopia. Nature 282, 29-33.

Williams, M.A.J., Williams, F.M. and Bishop, P.M. (1981). Late Quaternary history of Lake Besaka, Ethiopia. Palaeoecology of Africa 13, 93-104.

Chapter 9. Back to the Sudan: White Nile Valley and Jebel Marra Volcano

Adamson, D.A., Clark, J.D. and Williams, M.A.J. (1974). Barbed bone points from Central Sudan and the age of the "Early Khartoum' tradition. Nature 249, 120-123.
Adamson, D.A., Gasse, F., Street, F.A. and Williams, M.A.J. (1980). Late Quaternary history of the Nile. Nature 287, 50-55.
Adamson, D.A., Gillespie, R. and Williams, M.A.J. (1982). Palaeogeography of the Gezira and of the lower Blue and White Nile valleys. In: A Land between Two Niles, op.cit., 165-219.
Adamson, D.A., Clark, J.D. and Williams, M.A.J. (1987). Pottery tempered with sponge from the White Nile, Sudan. African Archaeological Review 5, 115-127.
Adamson, D., McEvedy, R. and Williams, M.A.J. (1993). Tectonic inheritance in the Nile basin and adjacent areas. Israel Journal of Earth Sciences 41, 75-85 (Ran Gerson Memorial Volume).
Ayliffe, D., Williams, M.A.J. and Sheldon, F. (1996). Stable carbon and oxygen isotopic composition of early- Holocene gastropods from Wadi Mansurub, north-central Sudan. The Holocene 6(2), 157-169.
Mawson, R. and Williams, M.A.J. (1984). A wetter climate in eastern Sudan 2,000 years ago? Nature 308, 49-51.
Obeid Mubarak, M., Bari, E.A., Wickens, G.E. and Williams, M.A.J. (1982). The vegetation of the central Sudan. In: A Land between Two Niles, op.cit., 143-64.
Philibert, A., Tibby, J. and Williams, M. (2010). A Middle Pleistocene diatomite from the western piedmont of Jebel Marra, Darfur, western Sudan, and its hydrological and climatic significance. Quaternary International 216, 145-150.
Talbot, M.R., Williams, M.A.J. and Adamson, D.A. (2000). Strontium isotopic evidence for Late Pleistocene re-establishment of an integrated Nile drainage network. Geology 28 (4), 343-346.
Williams, M.A.J. and Adamson, D.A. (1973). The physiography of the central Sudan. Geographical Journal 139(3), 498-508.
Williams, M.A.J. and Adamson, D.A. (1976). The origins of the soils between the Blue and White Nile rivers, central Sudan, with some agricultural and climatological implications. Occasional Paper No. 6, Economic and Social Research Council, National Council for Research, Khartoum, 39 pp, 12 figs.
Williams, M.A.J. and Adamson, D.A., eds (1982). A Land between two Niles: Quaternary geology and biology of the central Sudan. Balkema, Rotterdam, 246 pp.
Williams, M.A.J. and Clark, J.D. (1976). Prehistory and Quaternary environments in central Sudan. Palaeoecology of Africa 9, 52-3.
Williams, M.A.J. and Faure, H., eds (1980). The Sahara and the Nile: Quaternary environments and prehistoric occupation in northern Africa. Balkema, Rotterdam, 607 pp.
Williams, M.A.J. and Adamson, D.A. (1980). Late Quaternary depositional history of the Blue and White Nile rivers in central Sudan. In: The Sahara and The Nile. op. cit., 281-304.
Williams, M.A.J. and Williams, F.M. (1980). Evolution of the Nile Basin. In: The Sahara and the Nile, op. cit., 207-224.
Williams, M. and Nottage, J. (2006). Impact of extreme rainfall in the central Sudan during 1999 as a partial analogue for reconstructing early Holocene prehistoric environments. Quaternary International 150(1), 82-94.
Williams, M.A.J., Medani, A.H., Talent, J.A. and Mawson, R.A. (1974). A note on Upper Quaternary sub-fossil mollusca west of Jebel Aulia. Sudan Notes and Records 54, 168-172.
Williams, M.A.J., Clark, J.D., Adamson, D.A. and Gillespie, R. (1975). Recent Quaternary research in central Sudan. Bulletin de l'ASEQUA 46, 75-86.
Williams, M.A.J., Adamson, D.A. and Abdulla H. H. (1982). Landforms and soils of the Gezira: A Quaternary legacy of the Blue and White Nile rivers. In: A Land between Two Niles, op. cit., 111-142.

Williams, M.A.J., Adamson, D.A., Williams, F.M., Morton, W.H. and Parry, D.E. (1980). Jebel Marra volcano: A link between the Nile Valley, the Sahara and Central Africa. In: The Sahara and the Nile, 305-337.
Williams, M.A.J., Adamson, D.A. and Baxter, J.T. (1986). Late Quaternary environments in the Nile and Darling basins. Australian Geographical Studies 24, 128-144.
Williams, M.A.J., Adamson, D., Cock, B. and McEvedy, R. (2000). Late Quaternary environments in the White Nile region, Sudan. Global and Planetary Change 26, 305-316.
Williams, M.A.J., Adamson, D., Prescott, J.R. and Williams, F.M. (2003). New light on the age of the White Nile. Geology 31, 1001-1004.
Williams, M., Talbot, M., Aharon, P., Abdl Salaam, Y., Williams, F. and Brendeland, K.I. (2006). Abrupt return of the summer monsoon 15, 000 years ago: new supporting evidence from the lower White Nile valley and Lake Albert. Quaternary Science Reviews 25, 2651-2665.

Chapter 10. Wadi Azaouak, Niger

Talbot, M.R. and Williams, M.A.J. (1978). Erosion of fixed dunes in the Sahel, Central Niger. Earth Surface Processes 3, 107-113.
Talbot, M.R. and Williams, M.A.J. (1979) Cyclic alluvial fan sedimentation on the flanks of fixed dunes, Janjari, central Niger. Catena 6, 43-62.

Chapter 11. Petra and Wadi Rum, Jordan

Lawrence, T.E. (1935). Seven Pillars of Wisdom. Reprinted 2000, Folio Society, London, 384 pp.

Chapter 12. Algeria, Tunisia, and the Sahara

Rognon, P. and Williams, M.A.J. (1977). Late Quaternary climatic changes in Australia and North Africa: a preliminary interpretation. Palaeogeography, Palaeoclimatology, Palaeoecology 21, 285-327.
Williams, M.A.J. (1984). Geology. In: J.L. Cloudsley-Thompson (ed.), Key Environments: Sahara Desert. Oxford, Pergamon Press, pp. 31-39.
Williams, M.A.J. (1984). Late Quaternary prehistoric environments in the Sahara. In: J.D. Clark and S.A. Brandt (eds.), From Hunters to Farmers: The Causes and Consequences of Food Production in Africa, Berkeley, University of California Press, 74-83.
Williams, M.A.J. (1988). After the deluge: The Neolithic landscape in North Africa. In: J. Bower and D. Lubell (eds.), Prehistoric Cultures and Environments in the Late Quaternary of Africa. Cambridge Monographs in African Archaeology 26, BAR International Series 405, 43-60.
Williams, M.A.J. (1982). Quaternary environments in Northern Africa. In: A Land between Two Niles, op. cit., 13-22.

Chapter 13. Son and Belan Valleys, India, and Himalayan foothills

Biswas, R.H., Williams, M.A.J., Raj, R., Juyal, N. & Singhvi, A.K. (2013). Methodological studies on luminescence dating of volcanic ashes. Quaternary Geochronology, 17, 14-25.

Clark, J.D. and Williams, M.A.J. (1986). Palaeoenvironments and prehistory in north central India: a preliminary report. In: J. Jacobson (ed.) Studies in the Archaeology of India and Pakistan, Oxford, New Delhi, 18-41.

Clark, J.D. and Williams, M.A.J. (1990). Prehistoric ecology, resource strategies and culture change in the Son valley, northern Madhya Pradesh, central India. Man and Environment 15, 13-24.

Pal, J.N., Williams, M.A.J., Jaiswal, M. and Singhvi, A.K. (2004). Infra Red Stimulated Luminescence ages for prehistoric cultures in the Son and Belan valleys, north central India. Journal of Interdisciplinary Studies in History and Archaeology 1 (2), 51-62.

Patnaik, R., Sahni, A., Cameron, D., Pillans, B., Chatrath, P., Simons, E., Williams, M. and Bibi, F. 2009. Ostrich-like eggshells from a 10.1 million-yr-old Miocene ape locality, Haritalyangar, Himachal Pradesh, India. Current Science 96 (11), 1485-1495.

Pillans, B., Williams, M., Cameron, D., Patnaik, R., Hogarth, J., Sahni, A., Sharma, J.C., Williams, F. and Bernor, R. (2005). Revised correlation of the Haritalyangar magnetostratigraphy, Indian Siwaliks: Implications for the age of the Miocene hominids Indopithecus and Sivapithecus, with a note on a new hominid tooth. Journal of Human Evolution 48, 507-515.

Shane, P., Westgate, J., Williams, M. and Korisettar, R. (1995). New geochemical evidence for the Youngest Toba Tuff in India. Quaternary Research 44, 200-204.

Shane, P., Westgate, J., Williams, M. and Korisettar, R. (1996). Reply to Comments by S. Mishra and S.N. Rajaguru on "New Geochemical Evidence for the Youngest Toba Tuff in India'. Quaternary Research 46 (3), 342-343.

van der Kaars, S., Williams, M.A.J., Bassinot, F., Guichard, F., Moreno, E. (2012). The influence of the 73 ka Toba super-eruption on the ecosystems of northern Sumatra as recorded in marine core BAR94-25. Quaternary International, 258, 45-53.

Westgate, J.A., Shane, P.A.R., Pearce, N.J.G., Perkins, W.T., Korisettar, R., Chesner, C.A., Williams, M.A.J. and Acharyya, S.K. (1998). All Toba tephra occurrences across peninsula India belong to 75 ka eruption. Quaternary Research 50, 107-112.

Williams, M.A.J. (1982). Alluvial stratigraphy and archaeology: three examples from prehistoric Africa, India and Australia. In: W. Ambrose and P. Duerden (eds.), Archaeometry: An Australasian Perspective. Australian National University Press, Canberra, 112-119.

Williams, M.A.J. (1985). Pleistocene aridity in tropical Africa, Australia and Asia, In: I. Douglas and T. Spencer (eds.), Environmental Change and Tropical Geomorphology, London, George Allen and Unwin, 219-233.

Williams, M. (2004). Environmental impacts of extreme events: The Toba mega-eruption, volcanic winter and the near demise of humans. Journal of Interdisciplinary Studies in History and Archaeology 1 (1), 118-120.

Williams, M. (2012). The Toba super-eruption: history of a debate. Quaternary International 258, 19-29.

Williams, M. (2012). Did the Toba super-eruption have an enduring effect? Insights from genetics, prehistoric archaeology, pollen analysis, stable isotope geochemistry, geomorphology, ice cores, and climate models. Quaternary International, 269, 87-93.

Williams, M.A.J. and Clarke, M.F. (1984). Late Quaternary environments in north central India. Nature 308, 633-635.

Williams, M.A.J. and Clarke, M.F. (1995). Quaternary geology and prehistoric environments in the Son and Belan Valleys, north-central India. Geological Society of India Memoir 32, 282-308.

Williams, M.A.J. and Royce, K. (1982). Quaternary geology of the middle Son valley, north central India: implications for prehistoric archaeology. Palaeogeography, Palaeoclimatology, Palaeoecology 38, 139-162.

Williams, M.A.J. and Royce, K. (1983). Alluvial history of the middle Son valley, north central India. In: G.R. Sharma and J.D. Clark (eds.), Palaeoenvironments and prehistory in the middle Son valley, Madhya Pradesh, north central India, 9-21, University of Allahabad.

Williams, M.A.J., Pal, J.N., Jaiswal, M. and Singhvi, A.K. (2006). River response to Quaternary climatic fluctuations: Evidence from the Son and Belan valleys, north central India. Quaternary Science Reviews 25, 2619-2631.

Williams, M.A.J., Ambrose, S.H., van der Kaars, S., Chattopadhyaya, U., Pal, J., Chauhan, P R. and Ruehlemann, C. (2009). Environmental impact of the 73 ka Toba super-eruption in South Asia. Palaeogeography, Palaeoclimatology, Palaeoecology, 284, 295-314

Williams, M.A.J., Ambrose, S.H., van der Kaars, S., Ruehlemann, C., Chattopadhyaya, U., Pal, J., Chauhan, P. (2010). Reply to the comment on 'Environmental impact of the 73 ka Toba super-eruption in South Asia' by Martin A. J. Williams, Stanley H. Ambrose, Sander van der Kaars, Carsten Ruehlemann, Umesh Chattopadhyaya, Jagannath Pal, Parth R. Chauhan [Palaeogeography, Palaeoclimatology, Palaeoecology 284 (2009), 295-314].

Chapter 14. Afar Hominids, Ethiopia

Adamson, D.A. and Williams, M.A.J. (1987). Geological setting of Pliocene rifting and deposition in the Afar Depression of Ethiopia. Journal of Human Evolution 16, 597-610.

Clark, J.D., Asfaw, B., Assefa, G., Harris, J.W.K., Kurashina, H., Walter, R.C., White, T.D. and Williams, M.A.J. (1984). Palaeoanthropological discoveries in the Middle Awash Valley, Ethiopia. Nature 307, 423-428.

Williams, M.A.J. (1985). On becoming human: geographical background to cultural evolution. 11th Griffith Taylor Memorial Lecture. Australian Geographer 17, 175-185.

Williams, M.A.J., Assefa, G. and Adamson, D.A. (1986). Depositional context of Plio-Pleistocene hominid-bearing formations in the Middle Awash Valley, southern Afar Rift, Ethiopia. In: L. Frostick, R. Renaut, I. Reid and J.J. Tiercelin (eds.), Sedimentation in the African Rifts, Geological Society Special Publication No. 25, 241-251.

Chapter 15. Rajasthan, India

Singhvi, A.K., Williams, M.A.J., Rajaguru, S.N., Misra, V.N., Chawla, S., Stokes, S., Chauhan, N., Francis, T., Ganjoo, R.K., Humphreys, G.S., (2010). A ~200 ka record of climatic change and dune activity in the Thar Desert, India. Quaternary Science Reviews, 29, 3095-3105.

Chapter 16. Somalia

Williams, M. (2014). Potential impacts of damming the Juba Valley, western Somalia: Insights from geomorphology and alluvial history. Geophysical Research Abstracts vol. 16, EGU2014-1302, 2014.

Chapter 17. Inner Mongolia, China: Dealing with Desertification

Adamson, D.A., Williams, M.A.J. and Baxter, J.T. (1987). Complex late Quaternary alluvial history in the Nile, Murray-Darling and Ganges basins: Three river systems presently linked to the Southern Oscillation. In: V. Gardiner (ed.), Proceedings of the First International Conference on Geomorphology, Manchester, September 1985, John Wiley & Sons, Chichester, Part II, 875-887.

Lawrie, K. and Williams, M. (2004). Improving salinity hazard predictions by factoring in a range of human impacts in the context of climate change. Proceedings of the Cooperative Research Centre for Landscape, Environment and Mineral Exploration (CRC LEME) Regional Regolith Symposia 2004, edited by I.C.Roach, Canberra, November 2004, pp.199-203 (ISBN 0-9756895-0-9). (CD-ROM:ISBN 0-9756895-1-7).

Whetton, P., Adamson, D.A. and Williams, M.A.J. (1990). Rainfall and river flow variability in Africa, Australia and East Asia linked to El Niño - Southern Oscillation events. In: P. Bishop (ed) Lessons for Human Survival: Nature's record from the Quaternary. Geological Society of Australia, Symposium Proceedings 1, 71-82.

Williams, M.A.J. (1986). The creeping desert: what can be done? Current Affairs Bulletin 63, 24-31.

Williams, M.A.J. (1990). Desertification: human mismanagement or natural hazard? Geography Bulletin 22, 129-141.

Williams, M.A.J. (1995). Interactions of desertification and climate: present understanding and future research imperatives. Proceedings of the International Planning Workshop for a Desert Margins Initiative. Nairobi, January 1995, 161-169. Reprinted in Arid Lands Newsletter (2001).

Williams, M.A.J. (1995). Drought, desertification and climatic change. Proceedings of the International Scientific Conference on the Taklimakan Desert. Urumqi, China, September 1993. Arid Zone Research Supplement, 237-242.

Williams, M.A.J. (1997). Environmental futures and sustainable shares. Current Affairs Bulletin 74(2), 4-12.

Williams, M.A.J. (1999). Desertification and sustainable development in Africa, Asia and Australia. Proceedings, International Conference on Desertification and Soil Degradation, Moscow, November 11-15, 1999, pp. 107-124.

Williams, M.A.J. (2000). Desertification: general debates explored through local studies. Progress in Environmental Science 2 (3), 229-251.

Williams, M. (2004). Desertification in Africa, Asia and Australia: Human impact or climatic variability? Annals of Arid Zone 42, 213-230.

Williams, M. (2011). Desertification: Research imperatives. Annals of Arid Zone, 50, 1-9.

Williams, M.A.J. and Balling, R.C., Jr. (1995). Interactions of desertification and climate: An overview. Desertification Control Bulletin 26, 8-16.

Williams, M.A.J and Balling, R.C., Jr. (1996) Interactions of Desertification and Climate. Arnold, London, 270 pp.

Williams, M., McCarthy, M. and Pickup, G. (1995). Desertification, drought and landcare: Australia's role in an International Convention to Combat Desertification. Australian Geographer 26(1), 23-32.

Yang, Xiaoping and Williams, M. (2003). The ion chemistry of lakes and late Holocene desiccation in the Badain Jaran Desert, Inner Mongolia, China. Catena 51, 45-60.

Chapter 18. Flinders Ranges, Australia and Quaternary Climates in Australia

Byrne, M., Yeates, D.K., Joseph, L., Kearney, M., Bowler, J., Williams, M.A.J., Cooper, S., Donnellan,S.C., Keogh, J.S., Leys, R., Melville,J., Murphy, D.J., Porch, N., Wyrwoll, K.-H. (2008). Birth of a biome: insights into the assembly and maintenance of the Australian arid zone biota. Molecular Ecology 17, 4398-4417.

Chor, C., Nitschke, N. and Williams, M. (2003). Ice, wind and water: Late Quaternary valley-fills and aeolian dust deposits in arid South Australia. Proceedings of the Cooperative Research Centre for Landscape, Environment and Mineral Exploration (CRC LEME) Regional Regolith Symposia, edited by I.C.Roach, Adelaide, November 13-14, 2003, pp.70-73 (ISBN 0-7315-5221-0). (CD-ROM:ISBN 0-7315-

Cock, B.J., Williams, M.A.J. and Adamson, D.A. (1999). Pleistocene Lake Brachina: a preliminary stratigraphy and chronology of lacustrine sediments from the central Flinders Ranges, South Australia. Australian Journal of Earth Sciences 46, 61-69.

De Deckker, P., Kershaw, A.P. and Williams, M.A.J. (1988). Past environmental analogues. In: G. I. Pearman (ed.) Greenhouse: Planning for Climate Change. E.J. Brill, Leiden, 473-488.

Glasby, P., O'Flaherty, A. and Williams, M.A.J. (2010). A geospatial visualisation of a late Pleistocene fluvial wetland surface in the Flinders Ranges, South Australia. Geomorphology 118, 130-151.

Glasby, P., Williams, M.A.J., McKirdy, D., Symonds, R. and Chivas, A.R. (2007). Late Pleistocene environments in the Flinders Ranges, Australia: Preliminary evidence from microfossils and stable isotopes. Quaternary Australasia 24 (2), 19-28.

Haberlah, D., Glasby, P., Williams, M.A.J., Hill, S.M., Williams, F., Rhodes, E.J., Gostin,V., O'Flaherty, A., Jacobsen, G.E. (2010). 'Of droughts and flooding rains': an alluvial loess record from central South Australia spanning the last glacial cycle. In: P. Bishop and B. Pillans (eds), Australian Landscapes. Geological Society, London, Special Publications, 346, 185-223.

Haberlah, D., Williams, M.A.J., Halverson, G., McTainsh, G.H., Hill, S.M., Hrstka, T., Jaime, P., Butcher, A.R. and Glasby, P. (2010). Loess and floods: high-resolution multi-proxy data of Last Glacial Maximum (LGM) slackwater deposition in the Flinders Ranges, semi-arid South Australia. Quaternary Science Reviews 29, 2673-2693.

Mee, A., McKirdy, D.M., Krull, E.S. and Williams, M. (2004). Geochemical anlaysis of organic-rich lacustrine sediments as a tool for reconstructing Holocene environmental conditions along the Coorong coastal plain, southeastern Australia. Proceedings of the Cooperative Research Centre for Landscape, Environment and Mineral Exploration (CRC LEME) Regional Regolith Symposia 2004, edited by I.C.Roach, Canberra, November 2004, pp.247-251 (ISBN 0-9756895-0-9). (CD-ROM:ISBN 0-9756895-1-7).

Mee, A. C., McKirdy, D.M., Williams, M.A.J. and Krull, E.S. (2007). New radiocarbon dates from sapropels in three Holocene lakes of the Coorong coastal plain, southeastern Australia. Australian Journal of Earth Sciences 54 (6), 825-835.

Talbot, M.R., Holm, K. and Williams, M.A.J. (1994). Sedimentation in low gradient desert margin systems: a comparison of the late Triassic of north-west Somerset (England) and the late Quaternary of east-central Australia. Geological Society of America Special Paper 289, 97-117.

Walshe, K., Prescott, J., Williams, F. and Williams, M. (2001). Preliminary investigation of indigenous campsites in late Quaternary dunes, Port Augusta, South Australia. Australian Archaeology 52, 5-8.

Williams, M.A.J. (1984). Palaeoclimates and palaeoenvironments: (a) Quaternary environments. In: J.J. Veevers (ed.) Phanerozoic Earth History of Australia. Oxford, Clarendon Press, 42-47.

Williams, M.A.J. (1984). Cenozoic evolution of arid Australia. In: H.G. Cogger and E.E. Cameron (eds.), Arid Australia, Sydney, Australian Museum, pp. 59-78.

Williams, M.A.J. (1994). Some implications of past climatic changes in Australia. Transactions of the Royal Society of South Australia 118(1), 17-25.

Williams, M.A.J. (2000). Quaternary Australia: extremes in the Last Glacial-Interglacial cycle. In J.J Veevers (ed), Billion-year earth history of Australia and neighbours in Gondwanalan (ISBN 1 876315 04 0). GEMOC Press, Sydney, pp. 55-59.

Williams, M.A.J. (2001). Morphoclimatic maps at 18 ka, 9 ka, & 0 ka. In: J.J.Veevers, Atlas of Billion-year earth history of Australia and neighbours in Gondwanaland (ISBN 0-646-40974-30. GEMOC Press, Sydney, pp. 45-48. [Includes text and 5 colour maps/diagrams].

Williams, M.A.J. (2001). Quaternary climatic changes in Australia and their environmental effects. In: V.A.Gostin (ed), Gondwana to Greenhouse: Australian Environmental Geoscience (ISSN 0072-1085). Geological Society of Australia Special Publication No. 21, pp 3-11.
Williams, M. (2014). Interactions between fluvial and eolian geomorphic systems and processes: Examples from the Sahara and Australia. Catena, http://dx.doi.org/10.1016/j.catena.2014.09.015
Williams, M. A. J. and Adamson, D. A. (2008). A biophysical model for the formation of late Pleistocene valley-fills in the arid Flinders Ranges of South Australia. South Australian Geographical Journal 107, 1-14.
Williams, M. and Nitschke, N. (2005). Influence of wind-blown dust on landscape evolution in the Flinders Ranges, South Australia. South Australian Geographical Journal 104, 25-36.
Williams, M., Nitschke, N. and Chor, C. (2006). Complex geomorphic response to late Pleistocene climatic changes in the arid Flinders Ranges of South Australia. Géomorphologie: relief, processus, environnement, 4, 249-258.
Williams, M.A.J., De Deckker, P., Adamson, D.A. and Talbot, M.R. (1991). Episodic fluviatile, lacustrine and aeolian sedimentation in a later Quaternary desert margin system, central western New South Wales. In: Williams, M.A.J., De Deckker, P. and Kershaw, A.P., eds (1991). The Cainozoic in Australia: A re-appraisal of the evidence. Special Report No. 18, Geological Society of Australia, pp. 258-287.
Williams, M., Prescott, J. R., Chappell, J., Adamson, D., Cock, B., Walker, K. and Gell, P. (2001). The enigma of a late Pleistocene wetland in the Flinders Ranges, South Australia. Quaternary International 83-85, 129-144.
Williams, M., Cook, E., van der Kaars, S., Barrows, T., Shulmeister, J. and Kershaw, P. (2009). Glacial and deglacial climatic patterns in Australia and surrounding regions from 35 000 to 10 000 years ago reconstructed from terrestrial and near-shore proxy data. Quaternary Science Reviews 28, 2398-2419.

Chapter 19. Kenya

Ambrose, S.H., Hlusko, L.J., Kyule, D., Deino, A. and Williams, M. (2003). Lemudong'o: a new 6 Ma paleontological site near Narok, Kenya Rift Valley. Journal of Human Evolution 44, 737-742.
Ambrose, S.H., Bell, C.J., Bernor, R.L., Boisserie, J.R., Darwent, C.M., DeGusta, D., Deino, A., Garcia, N., Haile-Selassie, Y., Head, J.J., Howell, F.C., Kyule, M.D., Manthi, F.K., Mathu, E. M., Nyamai, C.M., Saegusa, H., Stidham, T.A., Williams, M.A.J. and Hlusko, L.J. (2007). The Paleoecology and paleogeographic context of Lemudong'o Locality 1, a Late Miocene terrestrial fossil site in Southern Kenya. Kirtlandia 56, 38-52.
Ambrose, S. H., Nyamai, C. M., Mathu, E. M. and Williams, M. A.J. (2007). Geology, geochemistry and stratigraphy of the Lemudong'o Formation, Kenya Rift Valley. setting. Kirtlandia 56, 53-64.

Chapter 20. Mauritania, and Southern Hemisphere

Gasse, F., Chalié, F., Vincens, A., Williams, M.A.J. and Williamson, D., (2008). Climatic patterns in equatorial and southern Africa from 30, 000 to 10, 000 years ago reconstructed from terrestrial and near-shore proxy data. Quaternary Science Reviews 27, 2316-2340.
Williams, M. (2004). Desertification in Africa, Asia and Australia: Human impact or climatic variability? Annals of Arid Zone 42, 213-230.

Williams, M., Cook, E., van der Kaars, S., Barrows, T., Shulmeister, J. and Kershaw, P. (2009). Glacial and deglacial climatic patterns in Australia and surrounding regions from 35 000 to 10 000 years ago reconstructed from terrestrial and near-shore proxy data. Quaternary Science Reviews 28, 2398-2419.

Chapter 21. Back to the Nile

Barrows, T.T., Williams, M.A.J., Mills, S. C., Duller, G.A.T., Fifield, L.K., Haberlah, D., Tims, S. G. and Williams, F.M. (2014). A White Nile megalake during the last interglacial period. Geology, 42, 163-166.
Honegger, M. and Williams, M. (2015). Human occupations and environmental changes in the Nile valley during the Holocene: the case of Kerma in Upper Nubia (Northern Sudan). Quaternary Science Reviews. (Accepted).
Macklin, M.G., Woodward, J.C., Welsby, D.A., Duller, G.A.T., Williams, F. and Williams, M.A. J. (2013). Reach-scale river dynamics moderate the impact of rapid Holocene climate change on floodwater farming in the desert Nile. Geology, 41, 695-698.
Macklin, M.G., Woodward, J.C., Toonen, W.H.J., Williams, M.A.J., Flaux, C., Marriner, N., Nicoll, K., Vertsaeten, G., Spencer, N. and Welsby, D. (2015). Re-evaluating river dynamics, hydroclimatic change and the archaeological record of the Nile Valley using meta-analysis of the Holocene fluvial archive. Quaternary Science Reviews. (Accepted)
Talbot, M.R. and Williams, M.A.J. (2009). Cenozoic evolution of the Nile basin. In: H. J. Dumont (ed). The Nile. Monographiae Biologicae 89, 37-60. Springer, Dordrecht.
Talbot, M.R., Williams, M.A.J. and Adamson, D.A. (2000). Strontium isotopic evidence for Late Pleistocene re-establishment of an integrated Nile drainage network. Geology 28 (4), 343-346.
Williams, M.A.J. (2009). Late Pleistocene and Holocene environments in the Nile basin. Global and Planetary Change 69, 1-15.
Williams, M.A.J. (2009). Human impact on the Nile basin: Past, present, future. In: H.Dumont (ed). The Nile. Monographiae Biologicae 89, 771-779. Springer, Dordrecht.
Williams, M. (2012). River sediments. In: C. Vita-Finzi (ed.), River History. Philosophical Transactions of the Royal Society of London, Series A, 370, 2093-2122.
Williams, M.A.J. (2013). Quaternary Geology and Environmental History of the Nile in the Sudan. Bulletin No. 42, 1-77, Geological Research Authority of the Sudan, Ministry of Energy & Mining, Khartoum, Sudan.
Williams, M. (2012). Geomorphology and Quaternary geology of the region between Wadi el Arab and Kerma. Université de Neuchâtel, Documents de la mission archéologique suisse au Soudan, 2012/4, 10-15.
Williams, M.A.J. and Talbot, M.R. (2009). Late Quaternary environments in the Nile basin. In: H. J. Dumont (ed). The Nile. Monographiae Biologicae 89, 61-72. Springer, Dordrecht.
Williams, M., Jacobsen, G.E. (2011). A wetter climate in the desert of northern Sudan 9900-7600 years ago. Sahara 22, 7-14.
Williams, M.A.J., Adamson, D., Prescott, J.R. and Williams, F.M. (2003). New light on the age of the White Nile. Geology 31, 1001-1004.
Williams, M., Talbot, M., Aharon, P., Abdl Salaam, Y., Williams, F. and Brendeland, K.I. (2006). Abrupt return of the summer monsoon 15, 000 years ago: new supporting evidence from the lower White Nile valley and Lake Albert. Quaternary Science Reviews 25, 2651-2665.
Williams, M.A.J, Williams, F.M., Duller, G.A.T., Munro. R.N., El Tom, O.A.M., Barrows, T.T., Macklin, M., Woodward, J., Talbot, M.R., Haberlah' D. & Fluin, J. (2010). Late Quaternary floods and droughts in the Nile Valley, Sudan: New evidence from optically stimulated luminescence and AMS radiocarbon dating. *Quaternary Science Reviews* 29, 1116-1137.

Williams, M.A.J., Usai, D., Salvatori, S., Williams, F.M., Zerboni, A., Maritan, L., Linseele, V. (2015) Late Quaternary environments and prehistoric occupation in the lower White Nile valley, central Sudan. Quaternary Science Reviews, http://dx.doi.org/10.1016/j.quascirev.2015.03.007

Williams, M.A.J., Duller, G.A.T., Williams, F.M., Macklin, M.G., Woodward, J.C., El Tom, O.A.M., Munro, R.N., El Hajaz, Y. and Barrows, T.T. (2015). Causal links between Nile floods and eastern Mediterranean sapropel formation during the past 250 kyr confirmed by OSL and radiocarbon dating of Blue and White Nile sediments. Quaternary Science Reviews http://dx.doi.org/10.1016/j.quascirev.2015.05.024

Woodward, J.C., Macklin, M.G., Krom, M.D. and Williams, M.A.J. (2007). The Nile: Evolution, Quaternary river environments and material fluxes. In: A.Gupta (ed.), Large Rivers: Geomorphology and Management. John Wiley & Sons, Chichester, pp. 261-292.

Woodward, J., Macklin, M., Fielding, L., Millar, I., Spencer, N., Welsby, D. and Williams, M. (2015). Shifting sediment sources in the world's longest river: the Sr isotope record of the Holocene floodplains in the desert Nile. Quaternary Science Reviews. (Under review).

Chapter 22. Epilogue

Byrne, M., Yeates, D.K., Joseph,L., Kearney, M., Bowler, J., Williams, M.A.J., Cooper, S., Donnellan,S.C., Keogh, J.S., Leys, R., Melville,J., Murphy, D.J., Porch, N., Wyrwoll, K.-H. (2008). Birth of a biome: insights into the assembly and maintenance of the Australian arid zone biota. Molecular Ecology 17, 4398-4417.

Place Index

M. Williams, *Nile Waters, Saharan Sands*, Springer Biographies,
DOI 10.1007/978-3-319-25445-6

Person Index

M. Williams, *Nile Waters, Saharan Sands*, Springer Biographies,
DOI 10.1007/978-3-319-25445-6

Subject Index

M. Williams, *Nile Waters, Saharan Sands*, Springer Biographies,
DOI 10.1007/978-3-319-25445-6

Printed in the United States
By Bookmasters